THE
DETERMINATION OF SEX

Gynandromorph Bulfinch. (From blocks given by Prof. Poll.) On the left, normal male: on the right, normal female. In the centre, gynandromorph with male coloration on the bird's right side, female on the left.

THE
DETERMINATION OF SEX

BY

L. DONCASTER, Sc.D.

FELLOW OF KING'S COLLEGE, CAMBRIDGE

Cambridge:
at the University Press
1914

CAMBRIDGE
UNIVERSITY PRESS

University Printing House, Cambridge CB2 8BS, United Kingdom

Cambridge University Press is part of the University of Cambridge.

It furthers the University's mission by disseminating knowledge in the pursuit of education, learning and research at the highest international levels of excellence.

www.cambridge.org
Information on this title: www.cambridge.org/9781107492844

© Cambridge University Press 1914

This publication is in copyright. Subject to statutory exception and to the provisions of relevant collective licensing agreements, no reproduction of any part may take place without the written permission of Cambridge University Press.

First published 1914
First paperback edition 2015

A catalogue record for this publication is available from the British Library

ISBN 978-1-107-49284-4 Paperback

Cambridge University Press has no responsibility for the persistence or accuracy of URLs for external or third-party internet websites referred to in this publication, and does not guarantee that any content on such websites is, or will remain, accurate or appropriate.

PREFACE

THIS little book was originally planned as a companion volume to my *Heredity in the Light of Recent Research*, which was first published in 1910. It was intended to treat more fully the problems shortly dealt with in a chapter added to the second edition of that book, and to show how recent investigations on Sex, approaching the subject from several different sides, tend to suggest conclusions in many respects in accord with one another. As the writing of the book progressed, it was found necessary to make considerable alterations in the original plan, in consequence of the differences in the condition of our knowledge of Heredity and of Sex.

The study of Sex has not yet reached a stage at which it is possible to give an account of established facts, and of generally accepted inferences from them, which shall be even comparatively free from controversial matter. The subject has been approached by many quite different lines, and these lines, although convergent, have as yet given no indisputable indication of the central point towards which they all tend. Under these circumstances, a completely impersonal survey of the facts, and of the interpretations which have been suggested, would be of little value, and consequently I have not hesitated to give my own opinion, to suggest interpretations which appear to me to be probable, or to criticize those of others. In this way I have tried to make the book rather more than a mere summary, and to give a general account of our present knowledge of the problem of sex-determination from the point of view of one engaged in research on the subject.

I have striven to avoid as far as possible the use of technicalities which might embarrass the untrained reader, but in a subject the roots of which ramify into almost every field of biology, the employment of some technical terms has been unavoidable. I have explained most of these where they are first used, and have defined the greater part of them in a glossary. A list of references has been added to all the chief authorities quoted, but in a book of this size it is impossible even to mention many very important contributions to the subject, and no attempt has been made either to give a complete bibliography, or to summarize the whole of our knowledge. My aim has rather been to discuss all the more important lines of evidence which bear on the problem of sex-determination, and to illustrate

Preface

each by one or more representative examples. The reader who should wish to follow any branch of the subject further, will easily find references to other works in those quoted in the list.

I have intentionally confined my account to the problems of sex-determination in animals, including Man, and have omitted all reference to the work of the same kind which has been done on plants. One reason for this is the limitation of space, but the problems raised by plants are in some respects different from those in animals, and it is perhaps preferable, at the present time at least, to consider them separately. When the subject of sex-determination in plants is more fully known, it will doubtless fall into line with the knowledge obtained from animals, and any complete account of the subject will have to include both, but for the present the omission does not leave a very serious blank.

Of the illustrations, about half a dozen are new, and are reproduced from drawings or photographs made for me by Mr E. Wilson; a few are reproduced, in some cases with some modifications, from well-known figures, and for the remainder I am indebted to the courtesy of the various authors for permission to copy their illustrations. In the case of all which are not original, the source is acknowledged in the description, and I take this opportunity of recording my sincere gratitude to the authors, editors and publishers, not only in England, but also in Germany and elsewhere, who have given me such valuable help in this matter. My thanks are more especially due to Prof. Poll, of Berlin, for giving me the blocks from which the frontispiece is printed, to Prof. Morgan, of New York, for lending original drawings of mutations in Drosophila (Plate VII), and to Mr H. Main, for giving me the photographs of the gynandromorphic ant and family of *A. betularia* reproduced on Plates XX and XXI.

L. DONCASTER.

CAMBRIDGE,
November 1914.

CONTENTS

CHAP.		PAGE
I.	THE PROBLEM	1

The universal distribution of sex, and our ignorance of its real function. The meaning of the expression 'determination of sex.' The facts which have to be explained.

| II. | THE NATURE AND FUNCTION OF SEX | 7 |

Definition of male and female—Ova and spermatozoa—Fertilization—Mingling of parental characters—Physiological differences between the sexes.

| III. | THE STAGE OF DEVELOPMENT AT WHICH SEX IS DETERMINED | 16 |

Possibility of sex-determination before, at, or after fertilization of the egg—Cases in which it appears to be determined by the egg apart from fertilization; *Phylloxera, Hydatina*, Gall-flies, etc.—Sex-limited transmission by the female—Sex-determination at fertilization; the Bee and other Hymenoptera—Polyembryony; *Litomastix*, human twins, Armadillo, 'free-martin'—Sex-limited transmission by the male.

| IV. | SEX-LIMITED INHERITANCE | 31 |

Explanation of the word 'sex-limited'—*Abraxas grossulariata*—Canary, Pigeon and Dove, Fowls—*Drosophila*—Human Cases; Colour-blindness, Night-blindness, Haemophilia—Yellow, black and tortoise-shell Cat—Exceptions—Abnormal sex-ratios.

CHAP.		PAGE
V.	THE MATERIAL BASIS OF SEX-DETERMINATION	50

The nucleus, chromosomes, and nuclear division—Maturation-divisions in the production of spermatozoa and ova—Fertilization—The chromosomes in Parthenogenesis; the Bee, Gall-fly, Aphids—Cases in which the male has one chromosome fewer than the female; 'X-chromosomes'—Idio-chromosomes—Cases in which the female has one chromosome less than the male; *Abraxas*—Chromosomes and sex-limited inheritance—Limitations of the hypothesis that chromosomes determine sex.

VI.	THE SEX-RATIO	73

Expectation that the sexes should be equally numerous—Conditions which affect the sex-ratio—Parthenogenesis—Breeding Seasons and physiological condition; Canaries, Dogs, Man—Age of parents—Difficulty of eliminating sources of error—Hertwig's experiments with Frogs; staleness of eggs; water-content of eggs—Sex-ratio also affected by male—'Indifferent' larvae—Effect of temperature on larvae—Abnormal sex-ratios among hybrids—Summary.

VII.	SECONDARY SEXUAL CHARACTERS	90

Meaning of the term—Possible causes of the differences between the sexes—Dependence of secondary sexual characters on the presence of functional genital organs—The hypothesis of hormones—Effects of castration and transplantation of genital organs in Mammals and Birds—Gynandromorphs in Birds prove the inadequacy of the hormone hypothesis—Castration and transplantation experiments in Insects—Physiological differences between male and female—'Parasitic castration' in Crustacea.

VIII.	THE HEREDITARY TRANSMISSION OF SECONDARY SEXUAL CHARACTERS	107

Evidence that each sex may transmit the secondary sexual characters proper to the other; experiments with Pheasants and Insects—Inheritance of horns in sheep—Inheritance in butterflies which have two or more forms of female—Hybrid moths in which the female has male characters—Hypothesis that sexual dimorphism may depend on a factor which has sex-limited inheritance.

CHAP.		PAGE
IX.	HERMAPHRODITISM AND GYNANDROMORPHISM . .	119

Occurrence of hermaphrodites — Hermaphrodite Nematodes and their chromosome-cycle—Crustacea; complemental males of Barnacles — Vertebrates—Gynandromorphs — Hypotheses of origin — Toyama's silkworms — Unisexual families; *Abraxas, Acraea*—Reciprocal crosses giving different results — Gynandromorphs as result of crossing.

X.	GENERAL CONCLUSIONS ON THE CAUSES WHICH DETERMINE SEX	136

Summary of evidence for sex-determination by means of two kinds of germ-cells—Difficulties to accepting this hypothesis as universal—Possibility that both eggs and spermatozoa are of two kinds, with selective fertilization—Objections to this hypothesis—Indications that sex depends on a physiological condition, which may be brought about by more than one cause.

XI.	THE DETERMINATION OF SEX IN MAN . . .	147

Difficulty of reconciling the belief that sex in Man may be influenced by the mother, with the observation that the spermatozoa are of two kinds—Evidence for the existence of male- and female-determining ova—Rumley Dawson's theory—Schenk's theory—Abnormal sex-ratios in families affected with sex-limited diseases.

GLOSSARY	159
LIST OF WORKS REFERRED TO	162
INDEX	168

LIST OF PLATES

Frontispiece. Gynandromorph Bullfinch.

PLATE		
I.	*Bonellia* and *Hydatina*	To face p. 5
II.	Life-cycle of the Gall-Fly (*Neuroterus*)	,, 18
III.	Polyembryony in Insects	,, 26
IV.	Quadruplet embryos in the Armadillo	,, 28
V.	*Abraxas grossulariata* (Currant Moth) and its variety *lacticolor*	,, 33
VI.	Sex-limited transmission of barred pattern in Fowls	,, 38
VII.	*Drosophila ampelophila* (Fruit-Fly)	,, 40
VIII.	Pedigrees of Sex-limited affections in Man (Colour-blindness and Haemophilia)	,, 43
IX.	Spermatogenesis of *Ascaris*	,, 53
X.	Maturation of the Egg in *Ascaris*	,, 54
XI.	Maturation and Fertilization of the Egg in *Cerebratulus*	,, 55
XII.	Spermatogenesis of the Hornet	,, 57
XIII.	Sex-chromosomes in Insects	,, 62
XIV.	Sex-chromosomes in the Eggs of Moths	,, 65
XV.	The Parasite *Sacculina* and its effects on its host	,, 102
XVI.	Hybrids between Moths of the genera *Biston* and *Nyssia*	,, 110
XVII.	*Papilio polytes* and its three forms of female	,, 113
XVIII.	Gynandromorphic Moths produced by crossing two species of the genus *Lymantria*	,, 115
XIX.	Chromosome-cycle in the Worm *Rhabdonema*	,, 121
XX.	Gynandromorphic Insects	,, 126
XXI.	Gynandromorphic family of the Moth *Amphidasys betularia*	,, 128
XXII.	Gynandromorphic Silkworm	,, 129

CHAPTER I

THE PROBLEM

THE question "Is it a boy or a girl?" is perhaps the first which is generally asked about the majority of mankind during the earliest hours of their independent existence; and the query "Will it be a boy or a girl?" must equally often be in the mind, even if it is less frequently expressed in words. This second question raises one of the most widely discussed problems of biology, that of the causes which determine whether any individual shall be male or female, and it suggests the still deeper question, "Why should there be male and female at all?" The problem of the nature and cause of Sex ranks in interest with that of the nature and origin of Life, and it may be that neither can be completely solved apart from the other. Notwithstanding the immense amount of brilliant speculation and research which has been devoted to the fundamental problem of Life, it must be admitted that hitherto no satisfactory solution has been found, and in some respects the question of Sex is equally obscure. Hardly any other problem has aroused so much speculation, and about few has there been such great divergence of opinion. In one direction, however, the last few years have seen a considerable advance, and we now know at least something of the causes which lead to the production of one or the other sex, although of the manner in which these causes act our ignorance is still profound.

It is but a short step from the question "Is it a boy or a girl?" to the further question "Why is it a girl instead of a boy?" and yet until recently the answer to the second question seemed hopelessly beyond our grasp, and even now, although some indications of an answer can be given, they do not touch the deeper problems of the real nature of sex. It is a remarkable thing that apart from the fundamental attributes of living matter—irritability, assimilation, growth and so forth—no single character is so widely distributed as sex; it occurs in some form in every large group of animals and plants, from the highest to the lowest, and yet of its true nature and meaning we have hardly a suspicion. Other widely distributed characters have obvious functions; of the real function of sex we know nothing, and in the rare cases where it seems to have disappeared, the organism thrives to all appearance just as well without it. And in many other cases, especially in plants, where sex is definitely present, it may apparently be almost or quite functionless, as, for example, in the considerable number of plants which are habitually grown from grafts or cuttings, and in which fertile seeds are never set. It is of course impossible to say with confidence that such "asexual" reproduction can go on quite indefinitely, but the evidence formerly adduced that continued vegetative reproduction leads to degeneration has been shown to be of doubtful validity. Sex, therefore, although it is almost universally found, cannot be said with certainty to be a necessary attribute of living things, and its real nature remains an apparently impenetrable mystery. A further problem, related to the last, arises from hermaphroditism—the presence of both sexes in one and the same individual. In the lower animals and many groups of plants this condition is frequently found, but its occurrence is quite irregular; it may be characteristic of large groups in some cases, of isolated species or even of individuals in others. It has sometimes been supposed that the hermaphrodite condition is primitive, and that the separation of the sexes into different individuals is a higher stage of evolution, but the evidence tends on the whole to show that the converse is true, and that hermaphrodite species have usually been derived from ancestors with separate sexes. The deeper problems of sex are still far beyond our comprehension, and up to the present time only one clear line of advance towards their solution has appeared, that of discovering, if possible, how the sex of the individual is determined. It is to this small branch of the subject that the present volume is devoted.

Before proceeding further, it may be well to explain more definitely what is meant by "the determination of sex." Popularly it is often supposed to mean the production of one or the other sex at will, but this is not the sense in which the phrase is used in biology. The study of the determination of sex is the study of the causes which lead to the production of an individual of one or the other sex, and those causes, when discovered, may or may not be amenable to human control. We may discover the causes of storms or earthquakes, and when our knowledge of them is sufficiently advanced we may perhaps be able to predict them as successfully as astronomers predict eclipses, but there is little hope that we shall ever be able to control them. So it may be with sex; a complete understanding of the causes which determine it may not necessarily give us the power of producing one or the other sex at will, or even of predicting the sex in any given case. Whether we shall ever be able to influence the causes of sex-determination cannot as yet be foretold; at present biologists are engaged in the less practical, but immensely interesting, problem, of discovering what those causes are. One other source of possible misunderstanding must also be referred to, that of the use of the word "cause" in connexion with sex-determination. If one condition is found to be followed invariably by a second, the first is called the (or a) cause of the second, but it does not follow that the one is the only or immediate cause of the other. The presence of a factor A may invariably be followed by condition E, but it may happen that A acts by setting up the chain of events B, C, D, and that any of these, if produced by a different "cause" would in turn lead to E. Hence, for example, it must not be assumed without further investigation that if the presence of a body A in an egg is always followed by the development of

Plate I

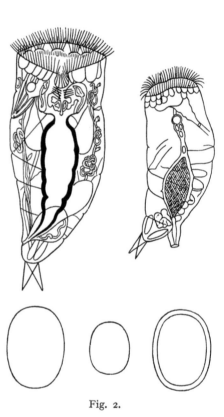

Fig. 2.

Fig. 1. *Bonellia viridis*, female, about half natural size. The four small crescent-shaped bodies to the right of the proboscis represent males drawn to the same scale. (From a specimen in the Cambridge Museum.)

Fig. 2. *Hydatina senta*, female on the left, male on the right. (Modified from Sedgwick after Cohn.) The relative sizes of the female egg, male egg, and fertilized "winter egg" are shown from left to right, on a larger scale, below.

Fig. 1.

that egg into a female, therefore all females are derived from eggs which contain A. They very probably may be, but that they are must be shown by independent proof.

With this introduction we may proceed to state the problem before us. Put shortly, it is to find explanations for the following facts. In the great majority of animals, and in many plants, every individual is either male or female. In each species there is a fairly constant average ratio between the number of males and females born; this ratio is commonly not far from equality, but varies considerably from species to species. In most animals and in some plants the difference between male and female does not concern only the organs directly connected with reproduction, but affects various parts of the body to a greater or less degree, so that the sexes are more or less readily recognisable by so-called "secondary sexual characters," such as the antlers of deer or the beard of man. Not rarely the differences are very striking, as in the peacock among birds, and finally we get the most extreme cases of "sexual dimorphism," as in the marine worm *Bonellia* (Plate I, fig. 1), in which the male is an almost microscopic parasite living in the body of the female, or in moths of the family *Psychidae*, and in some foreign species of *Orgyia*, in which the female is so degenerate as to be unable to leave the cocoon. These facts—the existence of two distinct sexes, the comparative regularity of the ratio in which they are produced, and the development of secondary sexual characters—are perhaps the most important branches of the subject, but they do not by any means exhaust the problems connected with the determination of sex. Hermaphroditism has already been mentioned, and it is obvious that any complete account of sex-determination must include an explanation of its cause. Another phenomenon which has very important bearings on the subject is Parthenogenesis, the production of offspring from eggs which have not been fertilized. Since in some cases such parthenogenetic eggs constantly give rise to males, in others to females, and in others again to either sex, it must be expected that valuable suggestions with regard to the mechanism of sex-determination should be obtained from a study of their behaviour. Finally, the relations between sex and heredity have given some of the most important clues, for if an animal transmits certain characters only to one sex among its offspring, it follows that the mechanism of sex-determination must be connected in some way with that of hereditary transmission.

These, then, are the chief problems which must be discussed, but it must be made clear at once that they are not all by any means solved at the present time. Our progress in the last few years has been rapid, but new difficulties appear as the old ones are removed, and a final answer can be given to hardly any of the questions raised. The purpose of this book is thus to show the present state of our knowledge, and to indicate the directions in which it seems that answers must be sought, rather than to set forth a completed theory of sex-determination.

CHAPTER II

THE NATURE AND FUNCTION OF SEX

An obvious preliminary to a discussion of sex-determination is a fairly clear idea of what is meant by the word Sex. In forms of life as various as mankind, birds, or insects, even in such trees as the willow and such Protozoa as *Vorticella*, we speak of some individuals as being male, and of others of the same species as being female. What is there in common between a hen and a willow-tree bearing green catkins that should justify our calling them female, while we say that the cock and the willow with yellow catkins are males? In the case of animals, a female may generally be defined as an individual which is capable of producing eggs, whether they are laid before development begins or whether the young develop to a greater or less extent before they are born. And it is no mere metaphor to describe a tree or plant, or a particular flower or part of a flower, as female, if it is a plant or part which bears seeds, for a seed is an embryo plant contained within an investing coat, just as an egg which has been incubated is an embryo chick within a shell. The fundamental thing, then, about the female sex is that female individuals produce bodies known as *egg-cells* or *ova*, which after uniting with a cell of a different character derived from the male, develop into new individuals. Superficially, egg-cells vary greatly in appearance; they may be relatively large, owing to the inclusion of nourishing substance or yolk for the developing embryo, or they may be microscopic, as they generally are when no yolk is present. They may have a special protective covering, or may be naked, but apart from these differences, which are so to speak accidental, they are always characterized, in the most various animals and plants, by consisting of a mass of relatively unmodified protoplasm containing a single nucleus[1].

As the distinguishing character of the female is the production of eggs or ova, so that of the male is the production of male germ-cells, which, however, vary greatly in different cases. They are characterized by the fact that their function is to reach an ovum and unite with it in the process of *fertilization*, as will be described in more detail below. In nearly all animals and in many of the lower plants, the male germ-cells are for this purpose endowed with the power of independent locomotion; in animals they are called *spermatozoa* (in the singular, spermatozoon, sometimes abbreviated to sperm) and in the lower plants *spermatozoids*. In the flowering plants the male germ-cells are enclosed in the

---

[1] Protoplasm is the name given to the substance which is the material basis of all living things. In chemical constitution it resembles white-of-egg, and consists of very complex compounds of carbon, hydrogen, oxygen and nitrogen, with a smaller amount of sulphur, phosphorus, and other mineral elements. Some account of the nucleus will be given below.

pollen-grains which are produced by the stamens of the flower; the pollen-grains have no power of independent movement, but are carried to the neighbourhood of the egg-cell in the female flower, or part of the flower, either by wind or by insects, and thence grow out a tube which penetrates to the egg-cell and carries the male germ-cell into contact with it. Although, therefore, there are great differences between the male germ-cells of different organisms, they all agree in one essential feature, they are adapted for reaching in some way the more stationary egg-cell, they unite with it in the process of fertilization, and the *zygote* so produced proceeds to develop into a new individual. In animals, in which the spermatozoa have the power of independent movement in a fluid, these are commonly more or less tadpole-shaped, consisting chiefly of a "head," which contains little else beside the nucleus of the cell from which they have been derived, and a vibratile protoplasmic tail by the motion of which they travel through the fluid in search of the egg-cell.

The essential feature of the process of fertilization is the union of the two nuclei contained respectively in the egg-cell and the head of the spermatozoon. The nucleus is a portion of protoplasm enclosed in a thin membrane and differing from ordinary protoplasm in containing a quantity of a substance called *chromatin,* so called because it takes up stains (Greek *chroma*) more readily than other parts of the cell. In its ordinary condition the chromatin is scattered in fine granules on protoplasmic threads enclosed in the nuclear membrane. We know that the nucleus is of fundamental importance to the life of the cell; metaphorically we may say that the nucleus is to the cell what the brain is to the body. When a spermatozoon meets an egg-cell, it forces its way into it, until its head is embedded in the egg-protoplasm; the tail is then dropped off, and the head, which consists almost entirely of a very concentrated nucleus, swells up until it reaches the size of the nucleus of the egg-cell (Plate XI, p. 55). The two nuclei then slowly approach each other until they meet, when they fuse into a single zygote-nucleus, which immediately begins to divide in such a way that equal parts of both parental nuclei are contained in each half. The whole cell then divides into two, and the process is repeated until an embryo consisting of thousands of cells is produced. In every division the nucleus is divided in such a way that both paternal and maternal portions are accurately halved; from this it results that every cell of the offspring includes equal parts of the maternal and paternal nuclear substance (chromatin). The details of the process of division will be considered in a later chapter, for by the study of them in the production of germ-cells and in the development of eggs more light has been thrown on the problem of sex-determination than perhaps by any other method.

In the process of fertilization we get almost our only definite indication of the ultimate nature and function of sex. We have seen that every part of every individual includes equal portions of nuclear matter descended from one and the other parent. The mechanism for producing this equal division is extremely beautiful and complex, and it is impossible to believe that it is not of fundamental importance. There is evidence that the nucleus, and particularly its chromatin, is especially concerned in the transmission of inherited characters, and the mechanism of nuclear division ensures that, of this "material basis of heredity," the portion derived from each parent shall be equally distributed to every part of the body. One of the chief effects, then, of sexual reproduction, and perhaps its most important function, is the equal mingling in every individual of sets of inherited characters derived from two parents. It would lead us too far to discuss what may be the advantages to the organism of this mingling in every generation of the parental qualities; it must suffice to point out that it occurs almost universally in plants and animals of every group.

It is obvious, however, that fertilization and the consequent mixture of inherited characters might occur in the absence of two distinct sexes. If all individuals produced similar germ-cells, and if these united with one another at random, we should still get fertilization and a similar recombination of characters, although there would be no division into males and females. How then has sex, as we know it, arisen? We do in fact find a condition practically identical with that suggested as possible, in some of the simplest animals and plants, the unicellular organisms known as Protozoa and Protophyta. In the common unicellular animal *Paramoecium*, for example, two similar individuals come into contact, the nucleus of each divides, and each transmits one portion to the other. The "migratory" nucleus of each individual unites with the "stationary" nucleus of the other; each has thus fertilized the other, and as the two are similar, neither can be called male or female[1]. In other unicellular organisms, two individuals, apparently quite similar to each other, unite completely; their nuclei fuse and they form a zygote which later divides into two fresh individuals. In the higher, multicellular animals, such union of similar cells is never found, and even in the Protozoa it is usual that the two cells which conjugate (*i.e.* unite in fertilization) should be dissimilar. One cause for this is not difficult to discover. The zygote formed by conjugation of two germ-cells has to grow up directly into a new individual, and if the function of fertilization referred to above is to be fulfilled, the germ-cells must come from different parents. If both cells were alike, it would be much more difficult for them to meet each other than if one is relatively large and stationary, the other small and active. Further, it is a great advantage to the embryo that it should be provided with some source of nourishment in its early stages. This is done in many cases by the storing up of food-material (yolk) in the egg, and it would clearly be impossible for two such yolk-laden cells to seek each other out and unite. In a number of animals belonging to various groups, and also in the flowering

[1] Some details of the process have been omitted, as not bearing on the question under discussion.

The Nature and Function of Sex

plants, the embryo is supplied with nourishment direct from the mother, and this again would be impossible if both germ-cells were alike. Hence it becomes clear that if once it is admitted that sexual reproduction is necessary or advantageous to the organism, the distinction of the two sexes, male and female, follows almost inevitably.

The facts mentioned in the last paragraph lead naturally to a consideration of the characteristics, apart from the production of eggs or spermatozoa, which distinguish the two sexes. It should be noted, however, that it is not certain whether the differences referred to follow from a primary division into egg-producing and sperm-producing individuals, or whether the differences themselves are really primary, and lead to egg-production in the female and sperm-production in the male as secondary consequences. This is a point of theoretical interest, to which reference will be made again in a later part of the book. Whether egg-production be primary, and the physiological characters which in general distinguish the sexes secondary, or the reverse, it is clear that there must in many cases be deep-seated physiological differences between male and female. From the point of view of the race, the function of the female is the production of eggs, and in the higher forms very frequently the nourishment and care of the young. The ova are relatively large passive cells, frequently laden with yolk, and their production involves the assimilation and building up of much food material, the storage of much matter and energy for the use of the next generation. The male germ-cell, on the contrary, is small and active, it contains a minimum amount of substance apart from its concentrated nucleus and vibratile tail, and instead of storing up energy for future use it rapidly uses up what it possesses in its vigorous search for the egg. It has often been pointed out that these differences are in general characteristic of the two sexes; the female of most species is relatively quiescent and devoted to the storing up of food and energy for the eggs and young, while the male is vigorous, restless and active, and characterized by the dissipation of energy rather than by the storing of it. In more technical language, the physiology of the female is relatively anabolic, that of the male catabolic in character. That this should be so is perhaps a necessary consequence of the difference in function between the sexes. The female cannot afford to dissipate, in needless activity, energy which is required for providing subsistence for the offspring; while the male, in many cases at least, has in consequence actively to search out the female, and not rarely has to fight with rivals for her possession. It is interesting to speculate, however, whether the active, vigorous habits of the male and the restless movement of the spermatozoon on the one hand, and the quieter habit of the female and the passivity of the egg on the other, may not each be due to fundamental catabolic and anabolic tendencies, characteristic of maleness and femaleness, quite apart from the exigencies of reproduction. That this may be so is suggested not only by the study of behaviour, but by the extraordinary development of apparently unnecessary ornament in some male birds and insects, and by the tendency of the female to degenerate in some cases to a mere egg-producing machine, in a way that would seem a danger to the survival of the species.

The other differences between the sexes are nearly always merely developments or corollaries of those already noticed. In the higher animals the female is usually less conspicuous, since her escape from enemies is less easy, and more important for the species, than in the case of the male. The male as a rule shows greater tendency towards new developments in structure or colour, while the female retains more of the ancestral characters of the race. Even in the mental qualities of human beings, these differences seem to appear, since woman is said to be more receptive and conservative, while men show greater originality and are more inclined towards change.

There are of course many exceptions to the rule that the male tends to vary more from the ancestral type than the female. When, however, one considers the enormous variety of secondary sexual ornament found in the males of families such as the Birds of Paradise, the Pheasants and the Ducks, or in the *Ornithoptera* group of butterflies, in which the females are all relatively similar, one cannot help thinking that whether these adornments are of use in securing mates or not, they must have originated in an inherent tendency on the part of the male toward such developments, and may then have been increased and perfected by sexual selection.

CHAPTER III

THE STAGE OF DEVELOPMENT AT WHICH SEX IS DETERMINED

In the adult animal there is no difficulty in deciding to which sex any individual belongs; even in species in which the sexes are externally similar an examination of the mature reproductive organs will always reveal the sex with certainty. In the early embryo, however, or sometimes even later, this is generally not the case; in most forms there are stages of development in which it is impossible to say whether the individual will become a male or a female. Hence arises the question, fundamental for our discussion, whether there is really a stage at which the individual is indifferent, and may according to circumstances become either male or female, or whether the sex is in reality irrevocably fixed from the beginning, although its visible signs do not appear until later. This is a problem about which there has been the greatest possible difference of opinion, and although in many cases a definite answer may now be given, it cannot be said with certainty that it is an answer which has universal application. If we consider an egg-cell, from its first origin in the ovary to its final development to a mature individual, we find that it passes through

Stage at which Sex is Determined

certain critical stages, any of which might be supposed to have importance in determining the sex of the individual to which the egg gives rise. An egg grows in the ovary to its full size and is discharged, it undergoes the process called maturation which will be described in Chapter V, its nucleus unites in fertilization with the nucleus introduced by a spermatozoon, and lastly it undergoes segmentation and development through embryonic stages to birth and maturity. Of these several periods, growth in the ovary, maturation, fertilization, and embryonic development, any one may conceivably be critical in producing one or the other sex, and according to the species chosen for investigation, evidence may be found that in any one of them is the decisive moment at which the sex is determined. Hence there have arisen at least three opinions as to the true period of sex-determination; the oldest was that sex was determined in embryonic life, while many more recent investigators have maintained that it is fixed either before fertilization, or by the act of fertilization itself.

When, however, it is remembered that there are only two sexes, and that in any species the average ratio in which they are produced is approximately constant, it seems unlikely that each of these conflicting opinions may be true in different cases, and it is therefore necessary to examine the evidence on which each depends, to see whether some sort of reconciliation cannot be found between them.

It will be most convenient at this stage first to consider cases in which sex appears to be fixed in the egg before it is laid, then those in which it is determined at fertilization, and to leave the evidence which indicates that sex may be influenced during embryonic or later life to a subsequent chapter. With regard to the two former conditions it will be impossible to present all the lines of evidence at this stage; only certain simple facts can be given which will lead up to a fuller discussion in later chapters.

The most obvious cases of the existence of definitely male- or female-producing eggs are those in which the eggs develop parthenogenetically, that is, without being fertilized, and in which it can be predicted with certainty before development begins whether the egg will give rise to a male or a female. This condition exists in insects belonging to at least two orders, and in the Rotifers ("wheel-animalcules"). It has long been known that in *Phylloxera*, the plant-louse which has caused terrible devastation among the vines of southern Europe, and in the Rotifer *Hydatina*, two kinds of eggs are laid, which differ from one another in size, and that of these the larger always produce females, the smaller, males. Here then we get a clear case of sex determined in

DESCRIPTION OF PLATE II.

Life-history of the Gallfly *Neuroterus lenticularis*. On the left, the parthenogenetic (spring) generation. (1) Fly with ovipositor withdrawn; (2) with ovipositor extended; (3) abdomen and ovipositor enlarged; (4) galls from which the parthenogenetic females hatch. On the right, the sexual (summer) generation, known as *Spathegaster baccarum*. (5) Male; (6) female; (7) abdomen and ovipositor of sexual female enlarged; (8) galls from which the sexual generation hatch. (The figures of the flies are from drawings by F. Balfour Browne, Esq.; the galls from specimens in the Cambridge Museum.)

Plate II

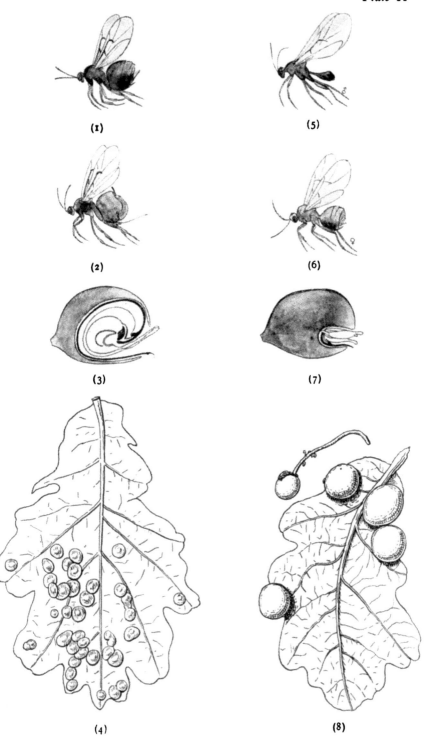

the egg during its growth; it should be noted, however, that it is conceivable that it is not the sex which is determined as such, for it might be argued that the presence of a larger amount of yolk in one case and a smaller amount in the other might so influence the embryonic development that the embryo with more yolk was converted into a female, that with less, to a male. Further investigation has shown that the two kinds of eggs are laid by different parthenogenetic mothers, so that some parthenogenetic females produce exclusively male offspring, others exclusively female. This distinction of parthenogenetically reproducing females into male-producers and female-producers has been found to be of wide, though not invariable, occurrence in the Aphids (plant-lice), in the Rotifers, and more recently in the Gallflies (*Hymenoptera*, Family *Cynipidae*). The life-cycle of the latter is remarkable (Plate II). There are commonly two generations in the year, the first of which consists entirely of females. Although there is no visible distinction into two classes, some of these females lay only male-producing, others only female-producing eggs, so that in the second generation both males and females are produced. These females pair with males and lay fertilized eggs, which give rise to the parthenogenetic females of the first generation[1]. One of the most remarkable points about this life-cycle is that the females of the first and second generations are so different from one

[1] The writer's experiments (1914) make it probable that individual sexual females have only male-producing or only female-producing parthenogenetic offspring (cf. p. 58 note).

another, that before their relationship was known, they were placed in different genera. The galls which the two generations produce on the oak-trees are also entirely unlike. In some species the sexual generation appears to be completely suppressed, and hence we find the very unusual condition of a species reproducing permanently by parthenogenesis; males are not known to exist, and fertilization appears never to take place.

An analogous, but not identical condition is found in the Rotifers (Plate I, p. 5). In them also there are male-producing and female-producing parthenogenetic females, which differ from each other only in the kind of egg which they produce. The males are small and degenerate; they pair indiscriminately with any female with which they come into contact, but no effect is produced on adult females, nor on female-producers at any stage. If, however, a male-producing female pairs when very young, the spermatozoa penetrate into the young eggs, and these, instead of growing into the small male-producing eggs, grow up into a third kind of egg, large, with thick shell, which can withstand drought or frost. These "winter eggs" when they hatch always yield females. In this case, therefore, as in the gall-flies, fertilized eggs always yield females, while parthenogenetic eggs are of two sorts, male-producing and female-producing.

In the Aphids the cycle is more varied, for there may be a number of generations of parthenogenetic females before sexual females and males are developed and there may be distinct male-producers and female-producers, or both males and females may arise from eggs of the same parent. As in the other cases, the fertilized eggs laid by the sexual females always give rise only to parthenogenetic females. These conditions will be discussed more fully when the nuclear phenomena accompanying sex-determination are described. The conditions found in the Aphids are closely paralleled by the Water-fleas of the family Daphniidae, as far as the life-cycles are concerned. The nuclear phenomena, however, of these little Crustacea have never been thoroughly investigated.

The last paragraphs may seem somewhat of a digression from the subject of sex-determination by the egg, but they illustrate the fact that some eggs may be male-producing, others female-producing, when not fertilized, and, in the Rotifers, that an egg which would have yielded a male when unfertilized, produces a female if it unites with a spermatozoon. This fact leads up to the subject of sex-determination by the act of fertilization, but one or two further points with regard to determination by the egg must be mentioned first. There are extremely few cases known in which two visibly distinct kinds of eggs are laid, one of which produces males and the other females, and both of which require fertilization. It was formerly supposed that this condition was found in the curious little marine worm *Dinophilus*, but the more recent work of Shearer has shown that in this animal the larger female-producing eggs conjugate with a sperm-nucleus, while the smaller male eggs do not. An example of eggs of two sizes, which give rise to different sexes, has been described by Reuter in the Mite *Pediculopsis*, and two kinds of eggs, differing in size, are known to exist in a Spider, but it has not been proved that those of one size always yield males, and those of the other, females. Since, therefore, visible determination of the sex in the egg before it leaves the ovary rarely occurs except in cases of parthenogenesis, and since in these same species fertilized eggs constantly yield females, it is open to those who believe that sex is determined by the act of fertilization to claim these cases as supporting their opinion rather than as disproving it. There is, however, one line of argument which, for the forms to which it is applicable, seems conclusive, but it is admittedly not of universal validity. A full account of it must be postponed to the next chapter, but for the sake of completeness it may be shortly mentioned here. It is found in certain moths and birds (*e.g.* the common Currant Moth and several breeds of fowls) that the female constantly transmits certain characters only to her male offspring. A hen with barred plumage, for instance, transmits the barred pattern only to her sons; her daughters, unless they receive the "factor" for barring from their father, are plain (Plate VI, p. 38). It seems clear from these facts that the barred hen must produce two kinds of eggs, some of which will develop into barred chicks and some into plain, and since all the barred chicks are males, and all the plain ones females, it seems equally clear that the eggs which transmit barring are male-producing, those which do not, female-producing. Since all the eggs require fertilization, we have here perfectly

Stage at which Sex is Determined

definite, though indirect, evidence, that distinct male-producing and female-producing eggs exist.

We must now return to the evidence for sex-determination at the moment of fertilization, to which some reference has been made already. The best known example is probably the Honey Bee. More than half a century ago Dzierzon put forward the hypothesis, which has ever since been known by his name, that, in the bee, queens and workers are produced from fertilized, drones from unfertilized eggs. He came to this conclusion from the fact that queens which have not paired, or which have exhausted the supply of spermatozoa received during pairing, produce only male offspring, and it was confirmed by the observation that a queen of one race, mated with a drone of another race, produces female offspring which show hybrid characters, but males which have only the characters of the mother's race. This last observation has been contradicted, but it is almost certain that it holds good except when the queen herself is not pure-bred. In recent years Dzierzon's theory has been abundantly confirmed by microscopic examination; eggs which are laid in worker or queen cells, that is, those eggs which will develop into females, contain spermatozoa, while those laid in drone cells do not. It seems, therefore, that a queen which has paired, and received a supply of spermatozoa into her seminal receptacle, can at will so to speak, allow an egg to be fertilized as it is laid, or withhold the spermatozoa so that it is not fertilized. It should be said that observations on the presence or absence of spermatozoa in the egg are supported by the character of the nuclei in the developing germ-cells of the two sexes, for males show conclusively that they possess only one complement of nuclear matter (chromatin), while females contain a double complement, half derived from one parent and half from the other.

Here then we get a perfectly definite instance, comparable with that already described in the Rotifers, of eggs which if unfertilized give rise to males, if fertilized, to females. All the eggs are as far as can be discovered originally alike; all undergo a similar maturation process, and the sex of the offspring seems to depend entirely on whether the egg-nucleus conjugates with a sperm-nucleus, or develops unfertilized.

The same phenomena that are found in the Bee are characteristic of nearly all insects of the order Hymenoptera. As will be seen more clearly in a later chapter, it appears to be universal in this order that the male has a single complement of chromatin in the nucleus, the female a double complement, and this condition is usually arrived at by unfertilized eggs developing into males, fertilized eggs into females. We have seen already, however, that in the Gallflies the female-producers lay unfertilized eggs which become females, and several other cases of a similar kind are known. In the parthenogenetic eggs of the Gallflies the behaviour of the nucleus is exactly what would be expected if the male is to have nuclei differing from those of the female in the same way as in other Hymenoptera (cf. p. 57), and an investigation of the nuclei of male and female pupae shows

Stage at which Sex is Determined

that the Gallflies are no exception to the ordinary rule for the order. In other cases, however, the behaviour of the egg-nucleus is less certainly known, and sometimes appears to be rather anomalous. In the Sawflies (Tenthredinidae) various conditions are found in different species. In the common Currant Sawfly, females which have not paired produce only male offspring (with, possibly, an occasional female, but this is doubtful), while females which have paired produce a large but variable percentage of females. This case is probably exactly analogous with that of the bee, the relatively few males produced by females which have paired being derived in all probability from eggs which have received no spermatozoon as they are laid. In some other Sawflies, however, both males and females, or in some cases only females, are produced by virgin mothers, and in some few species of the latter class, as in some Gallflies, no males are known to exist, and reproduction appears to be permanently parthenogenetic. The Sawflies are a group which have been much neglected from the point of view of sex-determination, and would probably well repay further study.

One of the most impressive examples of sex-determination at fertilization is provided by some of the parasitic Hymenoptera of the families Proctotrypidae and Chalcididae, which, like their relatives the Ichneumon flies, live in their larval stages inside the bodies of Caterpillars. Interesting examples have been described by Marchal (*Encyrtus*) and Silvestri (*Ageniaspis, Litomastix*). *Litomastix* (Plate III) is a minute insect which pierces the shell of the egg of the Silver Y Moth (*Plusia gamma*), and inserts its egg into the yolk. As the moth's egg develops into a caterpillar, the egg of the parasite develops within it. The mother fly lays with equal readiness whether she has paired or not. Her extremely minute egg contains a body of unknown origin[1] (the "paranucleus[2]"), which does not at first divide as the egg segments, so that it becomes enclosed in one of the cells into which the egg divides. Later on, however, this body breaks up into granules, some of which then pass into each cell when the cell containing them divides. When the egg has divided into a cluster of cells, this breaks up into a number of smaller clusters, and these, as they grow, may again break up in the same way. In this manner scores or hundreds of such cell-clusters are formed, some derived from cells which contained parts of the paranuclear body mentioned above, while others are without it. Each cluster now develops into a minute maggot-like larva, and these larvae feed on the fat and other less vital organs of the body of the caterpillar. The larvae derived from the groups of cells which contained no part of the paranuclear body develop to a certain stage and then stop. They

---

[1] While this was passing through the press, an important paper was published by R. W. Hegner (*Anat. Anz.* 46. 1914, p. 51), in which he describes the origin of this body from the nucleus of a cell of the ovary. This nucleus, instead of developing into an egg, becomes absorbed by a neighbouring egg and gives rise to the "paranuclear body."

[2] It should be noted that Marchal uses the word paranucleus in a sense different from that adopted here.

Plate III

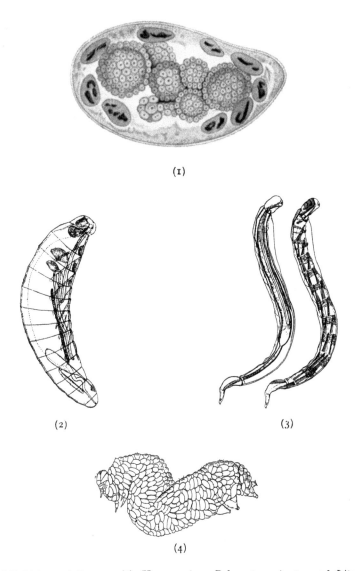

Life-history of the parasitic Hymenoptera, *Polygnotus minutus* and *Litomastix truncatellus*. (1) Embryo of *Polygnotus* dividing into secondary embryos (from Marchal). (2) Sexual, (3) sexless larvae of *Litomastix*. (4) Skin of host-caterpillar filled with pupae of *Litomastix*, all derived from one egg. (2-4 from Silvestri.)

Stage at which Sex is Determined

never show any trace of reproductive organs, and finally degenerate and die. The rest of the larvae continue their development until they are full-grown, by which time they have completely eaten up the body of the caterpillar; they then turn into pupae inside the empty skin of their host, and after a time emerge. It is sufficiently remarkable that only those larvae which are derived from cells containing part of the paranucleus should be able to form reproductive organs and to survive, but in connexion with the present subject the interesting thing about this life-history is that all the individuals produced from one egg, sometimes hundreds in number, are of one sex. If the original egg was fertilized, all are females; if the mother was virgin, all the young are males.

It appears to be a general rule that if an egg divides so as to produce more than one embryo, or even if it develops into a single embryo which later buds off sexual individuals, all the individuals produced from one egg are of the same sex. Interesting cases of this are known in certain mammals, including Man, and although a discussion of them is somewhat far removed from the subject of sex-determination at fertilization, it will be most convenient to deal with them at this stage. In Man it is a matter of common observation that twins may be of two kinds. Some twins are not more alike than other pairs of brothers and sisters, and such are often of different sexes. Other twins are so like each other as to be popularly called "identical twins," and these are always either both boys or both girls. Twins of the former class are derived from two ova which develop simultaneously, as in the litters of animals which produce several young at a birth. Identical twins are known, especially from the nature of their foetal membranes, to be derived from one ovum which has divided. Usually the division is complete, and thus two individuals are produced which have exactly or almost exactly the same hereditary constitution— hence their very close similarity. Less commonly the division is incomplete, and in that case double monsters arise, which range from two complete individuals only slightly united, as in the famous " Siamese Twins " to monstrous forms with two heads, or with one head but partially double body. The details of such division are outside the scope of the present subject; its importance in connexion with sex lies in the fact that when division occurs both individuals are always of the same sex.

The embryonic division which is abnormal in Man is the normal event in the Nine-banded Armadillo (*Tatusia novem-cincta*). In this species four young are regularly produced at a birth; they are all attached to a single placenta, and it has been shown conclusively by H. H. Newman that they are produced from a single ovum by embryonic division. In all the scores of Armadillo quadruplets which have been examined, they are always either four males or four females (Plate IV). This case thus provides striking confirmation of the already well-established belief that human identical twins arise from a single ovum. One curious fact, at present unexplained, must be mentioned in this

Plate IV

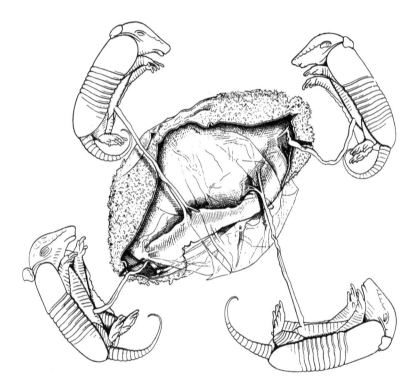

Nine-banded Armadillo (*Tatusia novemcincta*). Four foetuses attached to one placenta and derived from one ovum. (From a specimen presented to the Cambridge Museum by Dr H. H. Newman.)

III] *Stage at which Sex is Determined* 29

connexion. In cattle when a pair of twins is produced, one of which is male and the other apparently female, the latter may have abnormally developed genital organs and is known as a "free-martin." It seems probable that the free-martin is produced by the division of one ovum into two embryos, one of which gives rise to a normal male, the other to a sex-less individual which, from its lack of distinctively male features, is generally regarded as an abnormal and sterile female.

These last paragraphs have rather lost sight of the subject of the determination of sex at fertilization, but they at least deal with facts which show clearly that in many cases the sex must already be fixed immediately after the fertilization of the egg, since division of the early embryo always gives rise to individuals of like sex. Before leaving the subject, reference should be made to a group of facts bearing closely on the problem, which will be considered more fully in the next two chapters. It has been mentioned already that in some species of Moths and of Birds the female transmits certain characters only to her male offspring, and that it follows that two kinds of eggs, male-producing and female-producing, must exist. Exactly the converse is known in insects of other orders, and in Mammals (Cat, Man), in which the male transmits certain characters exclusively to his daughters. By the same reasoning as in the former case, it must be concluded that, in these animals, the male produces two kinds of spermatozoa, male-producing and female-producing, and, in fact, in a number of animals of different groups spermatozoa of two kinds, visibly distinct, are known, which correspond in their nuclear characters with the male and female respectively (cf. p. 62). Here, then, we seem to get definite evidence of sex-determination by the spermatozoon at fertilization; the only other possibility is that there are also two kinds of eggs, and that the male type of spermatozoon can fertilize only one kind of egg, the female type the other. Fuller discussion of this possibility must be postponed until the facts have been more completely described in the succeeding chapters, and it will be convenient also to leave any discussion of the possibility of sex-determination by influences acting in embryonic or later life until the facts referred to have been more adequately discussed

CHAPTER IV

SEX-LIMITED INHERITANCE

IN the previous chapter it has been mentioned that in some animals either the female transmits certain characters only to her sons, or, in other cases, the male transmits some characters only to his daughters. This peculiar type of inheritance was named by Bateson "sex-limited," to indicate that the transmission of the characters concerned is limited to one or the other sex. It has been objected that the term is misleading, and suggests that the character so transmitted is confined to one sex (as horns are confined to males in the Red Deer, or brown plumage to the hens of Pea-fowl); this is of course not the meaning, for characters which have sex-limited transmission may be present in either sex, as will be seen from the account which follows. To avoid this source of possible confusion, American writers on the subject have named this form of hereditary transmission "sex-linked inheritance," but, at least until our knowledge of the matter is more complete, the term "sex-linked" offers the serious objection of assuming that the "factor" for the character under consideration is linked with a sex-determining factor. It is highly probable that this hypothesis will prove correct, but in an argument which attempts to prove a certain conclusion, to use a term implying that conclusion in the premisses comes perilously near begging the question. It is therefore preferable to use words which involve no theoretical assumptions, and when it is understood that sex-limited inheritance has no necessary connexion with the inheritance of secondary sexual characters, there can be no objection to the use of the expression.

Sex-limited inheritance is now known in a considerable number of animals, in a Fly and a Moth among insects and in several Birds and Mammals; its existence is more than suspected in other animals, and almost proved in at least one plant. The first example to be thoroughly investigated, and the one in which it is perhaps most completely known in a quite simple form, is the common Currant Moth, *Abraxas grossulariata*. In the Currant Moth a very rare variety occurs, known as var. *lacticolor*[1]; it differs from the type, which in the following description will be designated *grossulariata*, or by the contraction *gross.*, in the reduction in the number and size of the black markings, and in their generally wedge-shaped outline, instead of the rounded outlines of the *grossulariata* markings (Plate V). *A. grossulariata* is a very variable insect, and in some (very rare) specimens the markings are reduced as much as

[1] According to the strict rules of priority in nomenclature, the variety should be known as *flavofasciata*. The name *lacticolor*, given by the Rev. G. H. Raynor, who first discovered some of the peculiarities of its inheritance, has come into such general use in the literature that it seems advisable to retain it.

Plate V

Abraxas grossulariata and its var. *lacticolor*. On the left, typical *A. grossulariata*, two males above, two females below. On the right, var. *lacticolor*, two males above two females below.

in a strongly marked *lacticolor*. Such lightly marked individuals might on mere inspection be regarded as intermediates, or even, if very carelessly examined, be confused with *lacticolor*, but the shape of the markings is always sufficient to separate the two forms. Among thousands which the writer has bred, he has never seen a specimen about which there was any doubt, and in at least 99 per cent. of those reared, the difference between the two forms is as striking as that between full-grown cock and hen pheasants.

The few *lacticolor* specimens which have been taken by collectors have all been females[1]. When a *lacticolor* female is mated with a wild *grossulariata* male, all the offspring, of both sexes, are *grossulariata*; this proves that the *lacticolor* character is a Mendelian recessive, that is, it lacks the "factor" for the *grossulariata* pattern, which is present in normal *grossulariata*. As will be described almost immediately, it is possible to obtain *lacticolor* males, and when one of these is paired with a wild *grossulariata* female, all the male offspring are *grossulariata*, all the females *lacticolor*. Reciprocal crosses thus give different results, which can only be interpreted by the assumption that a pure *grossulariata* male transmits the *grossulariata* pattern to his offspring of both sexes, while the *grossulariata* female transmits it only to her sons. Hence, as regards the transmission of the *grossulariata* pattern, the spermatozoa of the male are all alike, while there must be two kinds of eggs, those which transmit the *grossulariata* pattern

[1] There is one doubtful record of the capture of a wild male.

and develop into males, and those which do not transmit it, and yield females. The results of other matings completely confirm this hypothesis.

All the possible types of matings in which the *lacticolor* character can be involved have been made many times, and the results of them are given in the following list. In the symbolic representation of the factors concerned, G stands for the *grossulariata* factor, g for its absence, *i.e.* the *lacticolor* factor.

(1) *Lact.* ♀ × homozygous[1] *gross.* ♂ gives *gross.* ♂, *gross.* ♀.
    gg ♀   × GG ♂   „ Gg ♂, Gg ♀.
(2) *Gross.* ♀ × heterozygous[1] *gross.* ♂ gives *gross.* ♂, *gross.* ♀, *lact.* ♀.
    Gg ♀   × Gg ♂   „ GG ♂, Gg ♂, Gg ♀, gg ♀.
(3) *Lact.* ♀ × heterozygous *gross.* ♂ gives *gross.* ♂, *lact.* ♂, *gross.* ♀, *lact.* ♀.
    gg ♀   × Gg ♂   „ Gg ♂ gg ♂, Gg ♀ gg ♀.
(4) *Gross.* ♀ × *lact.* ♂   „   *gross.* ♂, *lact.* ♀.
    Gg ♀   × gg ♂   „ Gg ♂, gg ♀.
(5) *Lact.* ♀ × *lact.* ♂   „   *lact.* ♂, *lact.* ♀.
    gg ♀   × gg ♂   „ gg ♂, gg ♀.

The chief points of importance which follow from these results are firstly, that no homozygous *grossulariata* female can normally exist; in all of the many matings that have been made between *grossulariata* females and *lacticolor* males, *lacticolor* females have always resulted, so that it must of necessity be concluded that all *grossulariata* females have the constitution Gg, none GG. Normal (wild) *grossulariata* males, on the contrary, are homozygous for the *gross.* factor and have the constitution GG; they therefore transmit this factor to all their offspring of both sexes.

[1] As readers familiar with Mendelian heredity will know, the word "homozygous" means that the character concerned has been transmitted to the individual from both its parents; "heterozygous" that it has been transmitted by one parent but not by the other.

Sex-limited Inheritance

Secondly, although a heterozygous *grossulariata* male, *Gg*, transmits the *gross*. factor impartially to his offspring of both sexes, so that when mated with a *lacticolor* female half the male and half the female offspring are *grossulariata*, the heterozygous female transmits the *grossulariata* character to all her male offspring and to none of the females. There is thus no sex-limited transmission by the heterozygous male, and complete sex-limited transmission by the female. Owing to these facts, a *lacticolor* male can only be produced from a *lacticolor* female mated to a male which transmits *lacticolor*, for if the female parent is *grossulariata*, all her male offspring have the *grossulariata* factor. *Lacticolor* males are thus produced by only two types of mating—*lacticolor* ♀ × *lacticolor* ♂, which gives all offspring of both sexes *lacticolor*, since neither parent contains the *grossulariata* factor; and *lacticolor* ♀ by heterozygous *grossulariata* ♂ (*gg* ♀ × *Gg* ♂), in which half the offspring of each sex are *lacticolor* and half *grossulariata*.

As has been pointed out in the preceding chapter, a necessary deduction from these facts is that in this species there are two kinds of eggs, one male-producing and the other female-producing, and that the sex of the offspring must therefore be determined in the egg before fertilization. On no other hypothesis does it seem possible to account for the observed results with regard to hereditary transmission, which have been confirmed repeatedly in scores of experiments. One very important point, however, remains to be mentioned. In very rare cases the typical sex-limited transmission by the female fails, and a *grossulariata* female transmits the *grossulariata* factor exceptionally to a daughter. In *Abraxas* such failure of the sex-limited transmission is exceedingly rare; the writer has had only three *grossulariata* females among 611 female offspring of the mating *grossulariata* ♀ × *lacticolor* ♂ (in a total of 27 matings), and two of these were in one brood. In other animals, in which the same type of sex-limited inheritance occurs, such exceptions are less uncommon, and there is hardly any known example of sex-limited transmission that has been adequately investigated in which occasional exceptions do not occur. The general description of sex-limited inheritance given above must therefore be modified to this extent, that individuals of the sex in which sex-limited transmission occurs, normally or usually transmit certain characters exclusively to offspring of the other sex, but exceptions to this rule occur in nearly all known cases with greater or less rarity. These exceptions do not invalidate the conclusion that two kinds of eggs in the one case, or of spermatozoa in the other, are produced; they probably only show that male-determining eggs or female-determining spermatozoa do not invariably bear the sex-limited character. Another explanation of the facts is, however, conceivable, and will be considered at greater length in a later chapter. It is that in reality there are no exceptions to the rule that the sex-limited character is borne by ova or spermatozoa which determine the appropriate sex, but that in rare cases some influence acting at a later stage counteracts the sex-determining factor borne by the egg or spermatozoon, and thus changes the sex. It is conceivable, for example, that the very rare *grossulariata* females produced from the mating *grossulariata* ♀ × *lacticolor* ♂ may arise from male-determining eggs, which bear, as they should do, the *grossulariata* factor, but which have been made to develop into females by some cause acting at a later stage. This is doubtless an extremely improbable hypothesis; it is nevertheless a possibility which must not be rejected without due consideration.

Before turning to the second type of sex-limited transmission—that in which the male transmits certain characters only, or almost exclusively, to his daughters—some reference must be made to other cases of the *Abraxas* type which have been found in different animals. No other example in its simple form is known in Moths and Butterflies, but breeding experiments with the tropical *Papilios* (Swallow-tail Butterflies), which have two or more forms of the female and only one male form, are explicable only if it is assumed that one of the hereditary factors is sex-limited in the female. Similar explanations have been given of the behaviour of some other butterflies which have two distinct forms of female, so that it seems probable that sex-limited transmission by the female is not uncommon in Lepidoptera. In Birds it is known in at least three widely separated species. In the Canary, the Cinnamon variety has a pale brown pigment in the feathers, and deep red instead of black eyes, and this form behaves in relation with the black-eyed varieties just as the variety *lacticolor* behaves with respect to *grossulariata*. A pure-bred black-eyed male mated with a Cinnamon female gives all the offspring of both sexes black-eyed; a black-eyed female paired with a Cinnamon male has black-eyed male and red-eyed female young. The number of exceptions to the sex-limited transmission is, however, considerable, for among only 37 female young reared by Miss Durham from this mating, four had black eyes. In Pigeons and Doves, also, exceptions appear to be rather numerous; in Turtle-doves, Staples-Browne found that the transmission of the normal-coloured form is sex-limited when a coloured female is mated with a white male, but among 18 females reared one was coloured. In domestic Pigeons it has also been found that the factor for intense colour is sex-limited in transmission by the female, for when a full-coloured female is mated with a dilute-coloured male, the male offspring are full-coloured, the females dilute. A number of instances of sex-limited transmission by the female are known in Fowls. The first was worked out by Bateson and Punnett in crosses with the Silky breed, but their observations were complicated by the association of the sex-limited factor with a second factor which obscures the simple results. One of the clearest cases is that of the factor for barring in the Barred Plymouth Rock. When a barred cock is mated with an unbarred hen (Cornish Indian Game), all the chicks of both sexes are barred; when a barred hen of the same breed is mated with an unbarred cock, all the male chicks are barred, all the females plain (Plate VI). This result was first obtained by Pearl and Surface in America; it has been confirmed

Plate VI

Fig. 1. Fig. 2. Fig. 5. Fig. 6.

Fig. 3 Fig. 4. Fig. 7.

Results of crossing barred and unbarred breeds of Fowls, showing the sex-limited transmission by the female of the barred plumage. (By permission, from Pearl and Surface.) (1) Barred Plymouth Rock, male. (2) Cornish Indian Game, female. (3) Cornish Indian Game, male. (4) Barred Plymouth Rock, female. (5) Barred hybrid male. (6) Barred hybrid female, as produced from the cross Indian Game female (Fig. 2) × Plymouth Rock male (Fig. 1). (7) Unbarred hybrid female, as produced from the cross Plymouth Rock female (Fig. 4) × Indian Game male (Fig. 3).

by Goodale, and analogous experiments have been made by Davenport and others. Pearl has also made the discovery, important from the economic side, that the factor for high egg-production is sex-limited in the Fowl; it is therefore of no use simply to select hens from mothers with high egg-production, since the factor which determines this character is transmitted by the mother only to her male offspring. It follows that to get a race with high production the father of the selected hens should be the son of a hen with high egg-production, and only by mating such cocks with hens which themselves are good layers can a pure strain be built up. One further observation must also be referred to; Hagedoorn has reported that, in a breed of bantams, in one strain sex-limited transmission was found in the female, and in another strain the same character was sex-limited in transmission by the male. If this should be confirmed it would be of extreme importance; the account, however, is based on a single experiment, and as no other case of the kind is known, it must be accepted only with some reserve.

The second type of sex-limited inheritance, that in which the male transmits certain characters only to his daughters, was first known in Man and later in the Cat, but its nature was not clearly recognised until it was thoroughly investigated by T. H. Morgan in the Fruit-fly, *Drosophila ampelophila* (Plate VII). Morgan's series of papers on the inheritance of a considerable number of characters in this fly have now made it the classical example of this type of inheritance; many generations can be reared in a year, thousands of individuals can easily be bred, and under experimental conditions an extraordinary number of mutations have arisen, so that not only is the transmission of single sex-limited characters more completely known than in any other case, but also the relations between distinct sex-limited characters can be studied in a way that has not been possible in any other form. This work on the inter-relations of separate sex-limited characters in the same species is leading to results of fundamental importance for our knowledge of the mechanism of the transmission of hereditary characters, but it would involve too long a digression from the subject of the present volume to describe them here. It will be sufficient for the purpose of illustrating the close connexion between inheritance and the determination of sex to describe Morgan's experiments with one of the simpler characters, which give results entirely typical of all sex-limited cases in *Drosophila*. *Drosophila* is a little fly about half the length of a house-fly, and not unlike it in general appearance. Typically it has conspicuously red eyes, but in one of his earlier cultures Morgan found a few males with white eyes. These, mated with normal red-eyed females, gave over 1000 red-eyed individuals of both sexes, and in addition three white-eyed males which must be regarded as due to independent loss of the red-eye factor as in the first origin of the white-eyed form. The heterozygous red-eyed individuals mated together gave over 2000 red-eyed females, 1000 red-eyed males, and nearly 800 white-eyed males. A white-eyed male mated with a heterozygous red-

Plate VII

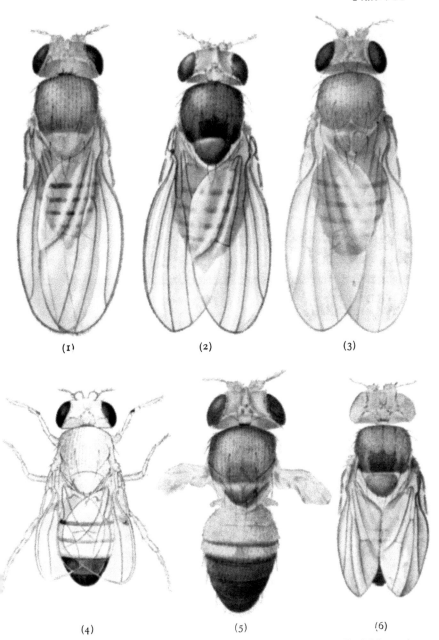

The Fruit Fly, *Drosophila ampelophila*. Forms which have arisen in Prof. Morgan's experiments. (1) Normal female; (2) Black female; (3) Yellow female; (4) Rudimentary-winged male; (5) Vestigial-winged male; (6) Black, white-eyed, miniature-winged male. All these mutations have sex-limited transmission by the male, except the black body and vestigial wing, and these two are "coupled" in their inheritance. (From drawings lent by Prof. T. H. Morgan.)

eyed female gave red-eyed and white-eyed males and females; the males and females in each class were almost exactly equal in number, but, as in the preceding mating, the reds of each sex were somewhat in excess of the whites in consequence of their greater constitutional vigour. Finally it was shown that white-eyed male and female gave nothing but white-eyed offspring, and that all red-eyed males, of whatever origin, when paired with white-eyed females, gave all female offspring red-eyed, all males white-eyed.

If R represents the red-eye factor, r the white-eye, the results are given diagrammatically thus:

(1) Red female × white male gives red males, red females.
    $RR\ ♀$ $rr\ ♂$ $Rr\ ♂$ $Rr\ ♀$.

(2) Heterozygous red female × red male
    $Rr\ ♀$ $Rr\ ♂$
    gives red males, white males, red females.
    $Rr\ ♂$ $rr\ ♂$ $Rr\ ♀, RR\ ♀$

(3) Heterozygous red female × white male
    $Rr\ ♀$ $rr\ ♂$
    gives red males, white males, red females, white females.
    $Rr\ ♂$ $rr\ ♂$ $Rr\ ♀$ $rr\ ♀$

(4) White female × red male gives white males, red females.
    $rr\ ♀$ $Rr\ ♂$ $rr\ ♂$ $Rr\ ♀$

(5) White female × white male gives white males, white females.
    $rr\ ♀$ $rr\ ♂$ $rr\ ♀$ $rr\ ♀$

It will be seen at once that these results are in every respect the exact converse of those obtained with *Abraxas grossulariata*. In *Abraxas* it has been shown that all *grossulariata* females are heterozygous for the *grossulariata* factor, and that they produce two kinds of eggs, of which the male-determining bear this factor and the female-determining do not.

In *Drosophila* precisely the same reasoning proves that the red-eyed male produces two kinds of spermatozoa, male-producing and female-producing, and that of these the female-producing bear the red-eye factor, the male-producing do not. While Morgan was experimenting with the white-eyed strain, new mutations appeared in his cultures; some of these involved other changes in the eye-colour, giving pink, vermilion, and orange eyes, others affected the body-colour, and still others the size and shape of the wings. Nearly all these new characters have been shown to be due to the loss of factors which normally exist in the typical fly, and in the great majority, when normals are paired with these new varieties, the normal characters are found to be sex-limited in transmission by the male, but never by the female. It is perfectly clear, therefore, that in *Drosophila* there are a number of characters—twenty or more have already been discovered by Morgan and his colleagues—which are regularly transmitted by the male only to his female offspring, while these same characters are transmitted by the normal female to all her offspring of both sexes, and if the female is heterozygous for one or more of them, she then transmits them impartially to half her sons and half her daughters. As far as the transmission of these characters is concerned, the eggs of *Drosophila* are all alike, while there are two separate kinds of spermatozoa, just as in *Abraxas* and the Fowl, Pigeon and Canary there are two kinds of eggs, but no evidence of more than one kind of spermatozoon. Clearly, then, it must be inferred, either that the sex

Plate VIII

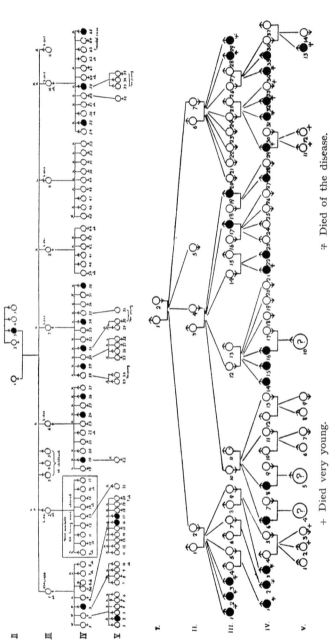

+ Died very young.   ‡ Died of the disease.

Pedigrees illustrating sex-limited inheritance in Man.
Above, pedigree of colour-blindness, showing transmission from an affected male, through his normal daughters, to some of his grandsons. (Slightly modified, in accordance with information received since its publication, from Nettleship, *Trans. Ophthal. Soc.* xxviii, p. 248.)
Below, pedigree of Haemophilia, illustrating transmission of the affection to males through normal females. The male No. 27 of generation IV is exceptional in apparently inheriting the disease direct from an affected father. (Reproduced, by permission of Prof. K. Pearson and Dr W. Bulloch, from *The Treasury of Human Inheritance*, Vol. I, Plate XXXVI, fig. 407, issued by the Francis Galton Laboratory for National Eugenics.)

is determined by the egg in *Abraxas* and the birds, and by the spermatozoon in *Drosophila*, or that there are two kinds both of eggs and spermatozoa in each case, but that in the former groups the sex-limited characters are associated with the differences in the eggs, in the latter group with those in the spermatozoa. It will be necessary to return to these alternatives at a later stage.

Although sex-limited transmission by the male is more thoroughly known in *Drosophila* than in any other animal, its existence was first recognised in Mammals, especially in the inheritance of certain abnormalities in Man. The commonest of these is colour-blindness; other abnormal conditions which appear to be inherited similarly are some forms of the affections of the eye known as Night-blindness and horizontal Nystagmus associated with deficient pigment, and especially the disease Haemophilia, the most conspicuous symptom of which is excessive bleeding from slight wounds. All these conditions occur more or less frequently in men, but are extremely rare in women. Affected men married to normal women have, as a rule, only normal children; their sons show no tendency to transmit the affection, but some of their sisters and probably all of their daughters may transmit it to their male children. It is thus commonly said that the diseases appear in men and are transmitted by women (Plate VIII). The explanation of this mode of inheritance, which seems so paradoxical at first sight, is quite simple. The affection arises through the loss of some factor which is present in the normal individual. The factor for normality may be represented by N, that for the disease by n (=the absence of N); the normal male is then heterozygous for N—just as the normal male *Drosophila* is heterozygous for the red-eye factor—and transmits it only by his female-determining spermatozoa. The normal female is homozygous for N, and thus transmits it by all her ova. The result of the mating normal by normal is thus shown in the following scheme:

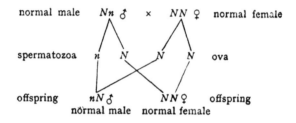

If now the factor N is missing in any male, he will have the constitution nn and will have the disease. Supposing him to marry a normal woman, we may represent his descendants thus:

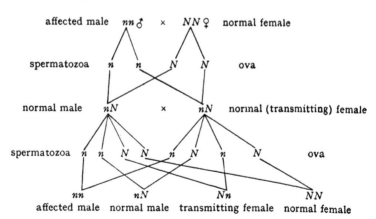

From this scheme it is seen that an affected man married to a normal woman has only normal children, and his sons, of constitution Nn, do not differ in any way from normal individuals. His daughters, however, although apparently normal, have the constitution Nn instead of NN; and since the factor N is not sex-limited in transmission by the female, they transmit N to some of their children of both sexes, n to the remainder. In consequence of the sex-limited transmission of N by the male, the normal man (Nn) transmits n to all his sons and N to all his daughters, and therefore some of the sons of a transmitting woman Nn will be affected (nn), and some normal (Nn); some of the daughters will be apparently normal but able to transmit (Nn), and the remainder completely normal (NN). The peculiar inheritance of the affections mentioned is thus fully explained on the assumption that the disease is due to the lack of something which is present in the normal, and that the factor for normality is sex-limited in transmission by the male, just as the white eye of *Drosophila* is due to the lack of the red-eye factor which is similarly transmitted. There is, however, one point of difficulty which will be mentioned after one other case of sex-limited transmission by the male has been described, that of the yellow, black and tortoiseshell colours in Cats. When a yellow male cat is mated with a black female, the male offspring are black and the females tortoiseshell; when a black male is mated with a yellow female, the female kittens are tortoiseshell as before, but the males yellow. These different results from reciprocal matings give clear indication of sex-limited inheritance, and the explanation appears to be that the factors for both black and yellow are sex-limited in the male and transmitted by him only to his daughters, while they are transmitted to both sexes by the female. In respect of these colours, the male can normally possess only one colour factor, while the female always has two. If Y represents the factor for yellow, B for black, and o the absence of colour factor, a yellow male is Yo, a yellow female YY, a black male Bo, a black female BB, and a tortoiseshell female BY. The matings mentioned would then be represented thus:

yellow male $Yo \times BB$ black female   black male $Bo \times YY$ yellow female

black male $oB \times YB$ tortoise female   yellow male $Yo \times BY$ tortoise female

This hypothesis completely accounts for the results usually obtained, but in rare cases tortoiseshell males are produced. To explain the existence of these it must be supposed that there is occasional failure of the sex-limited transmission by the male, resulting in his transmitting the factor Y or B to a male kitten instead of to a female. Since this male would at the same time receive either Y or B from its mother, it might then contain both Y and B and be a tortoiseshell. It is of interest that there is some evidence that such tortoiseshell males are generally sterile.

Such a failure of sex-limited transmission is exactly comparable to the exceptional cases already mentioned in connexion with sex-limitation in the female in *Abraxas*, the Canary and Turtle-dove. Exceptions of the same kind have been recorded by Morgan and his associates in *Drosophila*, and probably occur also in the inheritance of the human abnormalities. These exceptions will almost certainly prove to be of great importance in elucidating the true cause of sex-limited inheritance, for it is obvious that any explanation of the mechanism by which these characters are associated with the sex-determining factor not only must take into account the normal behaviour of the character, but must also be capable of explaining how such exceptions can arise.

A second problem which requires further investigation arises from the fact that although in general the results, both in the human cases and in the cat, are qualitatively exactly in accord with expectation on the assumption that the characters concerned are sex-limited, there are irregularities in the ratios which need further explanation. In the human cases, for example, if a woman who transmits colour-blindness or haemophilia marries a normal man, the theoretical expectation is that half her sons should be affected and half normal. It is generally found, however, in pedigrees of these human abnormalities not only that more than half the sons are affected, but also that there is an excess of sons over daughters. In *Drosophila* there is nearly always a deficiency of " affected " (white-eyed) individuals, and this can be accounted for by the known loss of vigour associated with the white eye. In human haemophilia pedigrees such an explanation is impossible, for there is almost constantly an excess of affected males, and associated with this, considerably more than half the children in affected fraternities are males. It has been suggested that both these numerical abnormalities are due to our incomplete knowledge of the pedigrees, for, even if every individual is recorded, any one family ("fraternity") can only be known to belong to the affected class if it includes at least one affected male. For this reason families consisting largely of daughters, and including no affected son, are excluded, and since such families would naturally balance those with a preponderance of sons, their exclusion from the totals would clearly tend to make the averages show an excess of males. This explanation is possibly the true one, but doubt is thrown upon it by the fact that very similar irregularities in the ratios are found in the Cat. If a tortoiseshell female is paired with a black male, the theoretical expectation is that half the male kittens should be yellow and half black; the writer found, however, in records of 22 matings obtained from breeders, that the ratio was 35 yellow to 29 black, and in the same litters there were altogether 67 males and only 35 females. In the converse mating in the human cases, that of an affected man married to a normal woman, there appears to be an excess of females among the offspring, and this clearly cannot be accounted for on the hypothesis suggested.

These facts indicate that possibly some cause hitherto not recognised may be at work in these cases, and further research will probably be required before the matter is fully explained.

Although there are some points which still remain obscure, the general conclusions to be drawn from sex-limited inheritance in relation to the problem of sex-determination are perfectly clear. From the facts described it seems impossible to doubt that sex-determining factors are borne by the ova and spermatozoa, and from the regularity of the observed results, that sex is in general fixed from the moment of fertilization and is not altered by events which may take place later. The important question remains whether sex is in some cases determined solely by the egg, the spermatozoon having no share in the process, and in other cases entirely by the spermatozoon with equally little participation by the egg. The facts of sex-limited inheritance alone would lead to this conclusion, and it gains some support from the observations on the behaviour of the nuclei which will be considered in the next chapter. It seems strange, however, that in one order of Insects and in the Birds among vertebrates the egg should be all-important as regards sex-determination, and that in other Insects and in Mammals sex should depend entirely on the spermatozoon. There are also other facts which make such a conclusion doubtful, and it will be seen at a later stage that another explanation is conceivable, although it involves difficulties which are hardly less than those which it is designed to avoid. This possibility amounts to assuming not that a single sex-determining factor is introduced sometimes by the egg and sometimes by the spermatozoon, but that sex is determined by the interaction of factors, of which one is derived from the male parent and the other from the female.

CHAPTER V

THE MATERIAL BASIS OF SEX-DETERMINATION

It is now necessary to turn to an entirely different aspect of the sex problem, an aspect which, nevertheless, has been found to be very closely connected with the subjects of the last two chapters. In discussing the sex of the offspring which arise parthenogenetically in the bee and other species it was mentioned that the constitution of the cell-nuclei in the male differs from that in the female, and that this difference provided the final proof that males in those cases are derived from unfertilized eggs. A more complete account of these facts was postponed until a general description should have been given of the behaviour of the nucleus in the origin and maturation of ova and spermatozoa, and in fertilization, after which it will be possible to discuss how this behaviour is related to sex-determination both in parthenogenesis and in the more usual case of bisexual reproduction.

The nucleus of any cell in its ordinary condition is enclosed in a membrane, and consists of a network of more solid substance bathed in a fluid. Scattered evenly over the network are exceedingly minute granules of the substance known as chromatin which is especially characteristic of the nucleus as

Material Basis of Sex-determination

distinguished from the cell-protoplasm. When the nucleus is about to divide the chromatin gradually collects into masses, probably by the contraction and concentration of the threads of the network. These chromatin masses are known as *chromosomes*, and in general both their number and their relative sizes are constant not only in all the cells of any individual, but in all the individuals of any species. Their number varies greatly in different species, so that the chromosome number may be regarded as a definite specific character.

After the chromosomes are fully formed, the nuclear membrane disappears and at the same time a spindle of threads arises, known as the division-spindle; the ends or points of the spindle are outside the nucleus and at opposite sides of it, and the threads run from these points or " poles " to the region that was occupied by the nucleus before the nuclear membrane disappeared (Plate XI, figs. 5-8). The widest part, or equator, of the spindle is in the region of the nucleus, and from the equator the threads converge to each pole. Some of the threads run direct from pole to pole; others become attached to the chromosomes. As the spindle is formed, the chromosomes come to lie in a circle or plane in the equator, and at this stage it can be seen that each chromosome is split, with a thread running from one pole to one half and from the other pole to the other half. It is at this stage that the chromosomes can most easily be counted, for if a view can be obtained of the circle or " equatorial plate " of chromosomes from the direction of one of the poles, since all the chromosomes lie in a flat plane they can all be focussed at once with the microscope, and can be counted without difficulty. As soon as the chromosomes have become arranged in the equatorial plane and have become definitely split, the halves of each move apart towards the two poles, possibly by the contraction of the spindle-fibre which connects each half with one of the poles. Thus every chromosome is divided into two exactly similar halves, and as the halves or "daughter-chromosomes" reach the poles, they swell up and form a nuclear network around which a nuclear membrane is formed. The original nucleus has thus divided into two daughter-nuclei each of which contains a complement of chromosomes exactly similar to that in the parent nucleus. The cell-protoplasm then divides in the plane of the equator of the spindle, and the process of cell-division is complete.

This, very shortly, is the method of cell-division in all cells of the body, at all stages except in certain divisions, known as "maturation-divisions," in the development of the ova and spermatozoa. A moment's consideration will show that it is impossible for the process described to go on unchanged from generation to generation if in fertilization both the ovum and the spermatozoon contained the same number of chromosomes as is characteristic of the species. Suppose, for example, the chromosome number to be 20. If the egg contained 20, and the spermatozoon also brought in 20, the fertilized egg would contain 40, and since at every cell-division each daughter-nucleus must then contain 40, in the next generation both egg and spermatozoon would

Plate IX

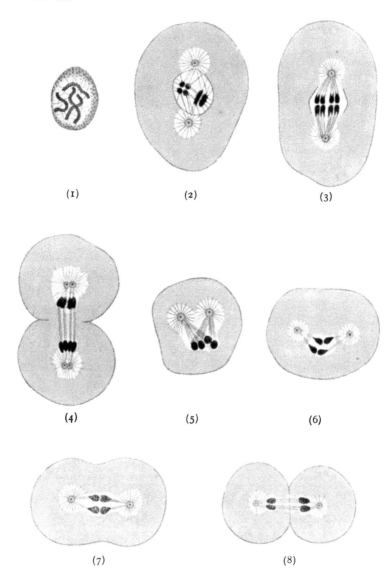

Spermatogenesis of the Round Worm, *Ascaris megalocephala*, after Brauer. (1) Nucleus of sperm mother-cell, with four chromosomes. (2) The chromosomes are arranged as two pairs; in the pair on the left each member has split, giving rise to a "tetrad." (3, 4) First maturation-division; separation of the members of each pair into separate daughter-cells. (5–8) Second maturation-division of one of the cells represented in fig. 4. The two split chromosomes divide so that two single chromosomes enter each daughter-cell. The cells represented in fig. 8 develop directly into spermatozoa.

contain 40, and the fertilized egg would then have 80. Such doubling of the chromosome number in each generation is obviously impossible, and the way by which it is avoided consists in the reduction of the chromosomes to half the number characteristic of the species in both ova and spermatozoa. It is not necessary for the purpose of the present discussion to describe in detail how this reduction is effected; in essence it consists of the union of the chromosomes in pairs before the " maturation divisions," and the subsequent separation of the members of each pair into the daughter nuclei at the next nuclear division. If the number characteristic of the species is 20, in the mother-cells of the spermatozoa the chromosomes unite into ten double chromosomes (Plate IX, in the species represented the normal chromosome number is four, the reduced number, two). The nuclear membrane then disappears, and these ten double chromosomes lie in the equator of the division-spindle. Each double then splits into its component single chromosomes, and thus ten complete chromosomes travel to each pole of the spindle, and are included in the daughter-nuclei when the cell divides. Since these ten are whole chromosomes, and not split halves, a second division immediately follows the first, in which each chromosome is split as in an ordinary division. It therefore follows that from each mother-cell four cells are produced, each containing a nucleus which has half the number of chromosomes normal in the species. Each of these cells is then converted directly into a spermatozoon; the cell-protoplasm gives rise to the tail and the nucleus becomes concentrated into the head. Every spermatozoon therefore contains half the number of chromosomes characteristic of the species.

In the maturation of the egg the process is essentially similar. It differs in the fact that the egg mother-cell, instead of giving rise to four eggs, gives rise to only one, and that three very small cells, containing hardly anything but a nucleus, are thrown off and lost. These little cells are called "polar bodies," since they are often formed at the pole of the egg which marks the anterior end of the embryo; their nuclei are spoken of as the polar nuclei. The maturation divisions of the egg commonly take place after it has been laid, and sometimes after the entrance of the spermatozoon, but always before the union of the egg-nucleus with the sperm-nucleus.

The maturation divisions usually take place as follows (cf. Plate X). The chromosomes unite in pairs at a very early stage in the ovary, before any yolk is laid down in the egg. After the egg has grown to its full size and is discharged from the ovary, the nucleus travels to one pole; its membrane disappears, a spindle is formed, and the double chromosomes come to lie in its equatorial plane. There they separate into their component single chromosomes, half of which travel to the outer pole of the spindle at the extreme edge of the egg, and half to the inner pole towards its centre. Immediately after they reach the poles, often without the formation of definite daughter nuclei, two new spindles are formed, each of which has in its equatorial plane one of the two groups of

Plate X

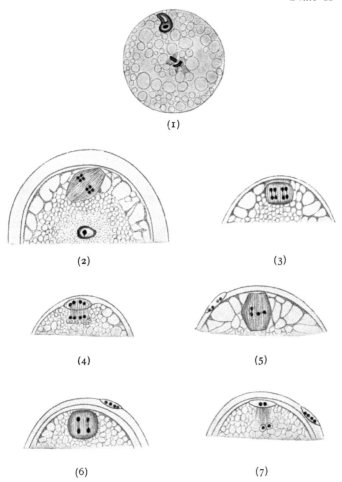

Maturation-divisions of the egg of the Round Worm, *Ascaris megalocephala* (from Korschelt and Heider, after Boveri). (1) Section of egg, showing egg-nucleus resolved into two double chromosomes in the centre, with spermatozoon at the upper edge. (2) First polar spindle at upper edge, spermatozoon in centre. The two components of each double chromosome have each split, producing two "tetrads." (3, 4) Division of tetrads into two pairs of "dyads," of which the outer two form the first polar body. (5–7) Second polar spindle, in which the two dyads left in the egg in fig. 4 each divide, leaving two single chromosomes in the egg, and two in the second polar body.

Plate XI

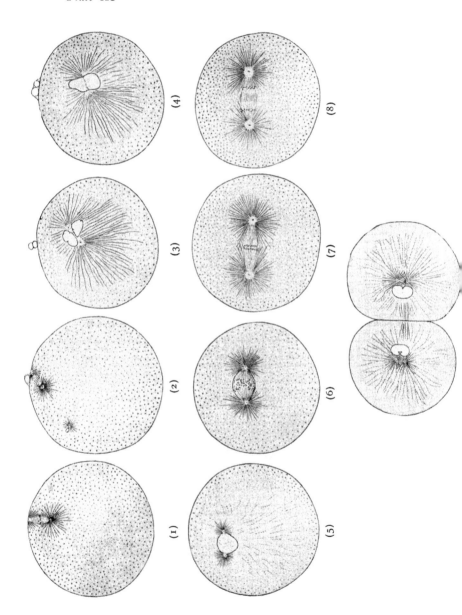

v] *Material Basis of Sex-determination* 55

chromosomes left at the poles of the first spindle, and each chromosome now splits in the normal manner and the halves travel to the poles of these second spindles. Four nuclei are thus formed, two at the poles of the outer spindle and two at the poles of the inner. The innermost of these four nuclei then sinks back into the centre of the egg and there unites with the sperm-nucleus; the three outer (polar) nuclei are either completely extruded from the egg as polar bodies, or in some cases simply disintegrate and disappear in the protoplasm of the egg margin. Although, therefore, the maturation processes of eggs and spermatozoa differ in detail, they are identical in the essential feature that the chromosome number of each is reduced to half the normal, and thus when the egg-nucleus and sperm-nucleus unite, the normal number is again restored. The differences may be regarded as being due simply to the necessity of the egg being relatively large, and the impossibility of its division into four equal cells when laden with yolk, while the spermatozoa have to be as small as possible, so that the production of four from one mother-cell is both possible and

DESCRIPTION OF PLATE XI.

Fertilization and first segmentation division of the egg of the worm *Cerebratulus* (after Coe). (1) First polar division. (2) Second polar division, with sperm-nucleus to the left. (3, 4) Conjugation of sperm-nucleus with egg-nucleus; the two polar bodies are seen at the upper edge of the egg. (5) The zygote-nucleus preparing to divide. (6) Formation of the division-spindle, and chromosomes arranging themselves on its equator. (7) Splitting of the chromosomes. (8) Split halves of the chromosomes travelling to the poles of the spindle. (9) Formation of the daughter-nuclei, and division of the cell.

economical. One point of great interest must be referred to before proceeding to discuss the relation of these phenomena to sex-determination. In many species the chromosomes differ among themselves considerably in size, and, with one exception to be described later, there are always two of each size in nuclei which have not undergone the maturation process. In the pairing which precedes maturation, it is always similar chromosomes which unite in pairs; if the chromosomes are lettered according to size, $A, B, C, \ldots$ there will be two A's, two B's, and so on, and in pairing A pairs with A, B with B, and C with C. Hence the egg-nucleus and sperm-nucleus, after the maturation divisions, each contain one complete set, $A, B, C, \ldots$, and when the nuclei unite in fertilization the zygote-nucleus produced by their union again contains the double complement $A, A, B, B, C, C, \ldots$

The facts described will make clear the statement made in Chapter III about the difference between the male and female bee. The male arises from an unfertilized egg which has undergone maturation; such an egg should have the reduced or half number of chromosomes, the single complement $A, B, C, \ldots$ The female is produced from a fertilized egg, and should therefore have the full or double number, $A, A, B, B, C, C, \ldots.$ The male (drone) has in fact 16 chromosomes, the female (queen) 32, in the genital cells before maturation. An obvious difficulty then arises; if the male already has the reduced number, how can normal maturation divisions take place in the development of the spermatozoa?

Plate XII

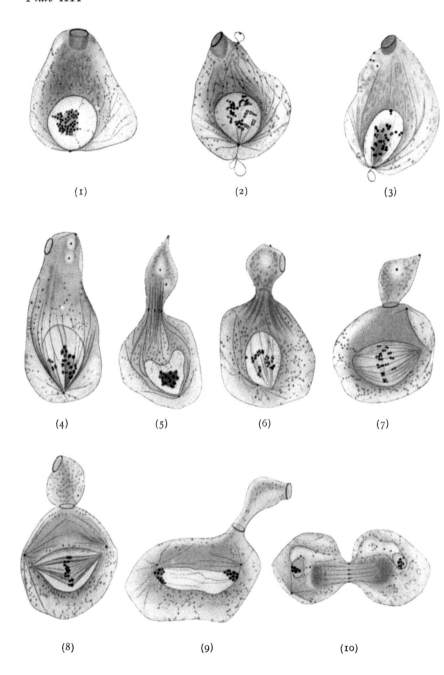

v] *Material Basis of Sex-determination* 57

The answer to this is that one of the maturation divisions is suppressed (Plate XII); each sperm-mother-cell divides only once, and since no pairing of chromosomes has taken place, the 16 chromosomes merely split as in an ordinary cell-division, and all the spermatozoa have 16, the normal reduced number. Here, then, is a clear case of sex determined by, or at least in connexion with, the presence of a definite number of chromosomes; when the full, or double, number is present, the individual is a female; when only the half number is present, it becomes a male.

In the bee the presence of the full or of the reduced number of chromosomes depends upon whether the egg is or is not fertilized. In the Gall-flies (Cynipidae), the same results are attained in another way. It will be remembered that many of these flies have two generations in the year, one of which consists entirely of parthenogenetic females, the other of males and females. The parthenogenetic females of the first generation are of two kinds, one of which lays male-producing eggs and the other female-producing. It is found that the

DESCRIPTION OF PLATE XII.

Spermatogenesis of the Hornet (by permission, from Meves). (1) Sperm mother-cell. (2–6) Abortive division of the nucleus, resulting in the separation of a piece of protoplasm above, but leaving the whole nucleus in the larger (lower) cell. (7–10) Normal division of the large cell into two daughter-cells, each of which gives rise to a spermatozoon. [Note. The spermatogenesis of the Hornet is figured rather than that of the Bee, because in the Honey-Bee there are peculiarities, not referred to in the text, which are not present in most other Hymenoptera.]

male-producing eggs undergo maturation and reduce the chromosome-number in the nucleus from 20 to ten. The eggs laid by female-producers undergo no maturation process; the nucleus comes to the surface as if to divide and give off polar nuclei, but undergoes no division, sinks back to the centre and begins its development into an embryo without further preparation. It contains 20 chromosomes, and therefore the females of the second generation have 20, the males ten chromosomes. As in the bee, in the developing spermatozoa one maturation division is suppressed, so that the spermatozoa have ten chromosomes; all the eggs of the second generation throw off polar nuclei and reduce their chromosomes to ten, and thus the fertilized eggs have 20 and give rise to the parthenogenetic females of the first generation. It is not known what determines whether these first generation females shall be male-producers or female-producers[1].

The Gallfly is a case of peculiar interest, for it shows that the origin of a male or a female from a parthenogenetic egg may depend on, or at least be absolutely correlated with, the presence or absence of a particular set of chromosomes. The condition found in the Hymenoptera (the order of insects to which the Bee and Gallfly belong), is however pecu-

[1] Recent experiments made by the writer in the hope of throwing light on this question make it probable that every individual sexual female has only male-producing or only female-producing parthenogenetic offspring. If this is confirmed, it would suggest that there may perhaps be differences, hitherto unrecognised, between the maturation-processes of the eggs laid by the two classes of sexual females.

liar; no other group of animals is known with certainty in which the male has only half as many chromosomes as the female. The same condition probably exists in the Rotifers, and has been suspected in certain Crustacea (the Phyllopoda, which include the "Water-fleas"); in this last case there is no good evidence for it, and in most other groups of animals it is known not to exist. A condition somewhat analogous to it, however, has been discovered by Morgan and von Baehr in Aphids (Plant-lice or Green-fly, including *Phylloxera*). As has been described in Chapter III, in many of these insects there are male-producing and female-producing parthenogenetic females, as in the Gallflies, and it has been found that in the eggs of both there is only one maturation division, giving rise to the egg-nucleus and one polar nucleus. There is no pairing of the chromosomes before this division, and in the female-producing eggs all the chromosomes split normally, so that the female has the full double number of chromosomes. In the male-producing eggs, on the other hand, one chromosome, instead of splitting, goes undivided to the outer pole of the spindle, and is therefore extruded with the polar nucleus; the egg-nucleus is thus left with an odd number of chromosomes, since the one that has been extruded has left its mate, which may be called X, unpaired. The female has the double number A, A, B, B, X, X, the male has A, A, B, B, X. In the maturation of the eggs laid by the sexually reproducing female these chromosomes unite in pairs and undergo the double maturation division, resulting in an egg-nucleus with the single chromosomes A, B, X; in the development of the spermatozoa A pairs with A, B with B, but the single X chromosome, having no mate, goes to one pole of the first division-spindle, and leaves the other pole with no X chromosome. If both the daughter cells of this division developed further, two kinds of spermatozoa would thus be produced, half having nuclei containing A, B, X and half having only A, B. The cell containing only A and B, however, never continues its development; it is smaller than its sister cell, and degenerates altogether, so that all the mature spermatozoa have the full single complement A, B, X. Since the eggs also all have the same complement of chromosomes, all fertilized eggs grow up into individuals having the full double complement A, A, B, B, X, X, and all these are females which again start the parthenogenetic cycle. The Aphids thus resemble the Bee and other Hymenoptera in having a different chromosome-total in the male and the female; they differ from them in the fact that the difference in the chromosomes concerns one chromosome only, instead of a whole single complement. If we assume that the difference between male and female consists in the presence of two X-chromosomes in the female and of one in the male, the two cases can be brought into line by supposing that one of the 16 chromosomes of the male Bee, or of the 10 chromosomes of the male Gallfly, is an X-chromosome, in which case there must be two X-chromosomes among the 32 of the female Bee or the 20 of the female Gallfly.

v] *Material Basis of Sex-determination* 61

The condition in which the male has one less chromosome than the female, or, in other words, in which the male has one X-chromosome and the female two, has been illustrated first by the example of the Aphids because they provide a connecting link between the better-known instances of parthenogenetic and of sexual reproduction. It has been shown that in them males arise from eggs from which one of the two X-chromosomes has been extruded; that since the male has only an unpaired X-chromosome, two kinds of spermatozoa begin to be formed; and that, since the cells without the X-chromosome degenerate and never form spermatozoa, all fertilized eggs have two X-chromosomes, and all develop into females. These facts are completely in accord with what has been discovered in other animals in which parthenogenesis does not take place, and if the historical order had been followed, it would have been necessary to describe these other cases first. It was known at the close of the last century that in insects of the order *Orthoptera* (Grasshoppers, etc.) one chromosome behaved differently from the others in the development of the spermatozoa, and it was soon found that this chromosome was unpaired, and that in consequence half the spermatozoa possessed it and half were without it. Its difference in behaviour consists in its remaining as a compact body while the other chromosomes have the form of elongated loops at the stage at which the pairing of the ordinary chromosomes takes place, and it is this difference in behaviour, in addition to its apparent connexion with sex-determination, that justifies its designation by a special symbol as the X-chromosome[1]. After the unpaired X-chromosome had been discovered in the males of certain *Orthoptera* and *Hemiptera* (Plantbugs), it was found that a pair of such chromosomes was present in the females of the same species (Plate XIII, figs. 1, 2). Since, therefore, in these forms, the female before the maturation divisions has two X-chromosomes and the male only one, it follows that after maturation all eggs possess an X-chromosome, while half the spermatozoa have it and half do not. The eggs which are fertilized by spermatozoa containing X will then give rise to individuals which have two X-chromosomes, and will become females, while those fertilized by spermatozoa without X will develop into individuals with only one X, and these will be males. A similar condition has now been found in insects belonging to several orders, and in a number of animals of other groups. If the condition described were found universally, the problem of sex-determination might be considered as almost solved; it would at least be possible to say that females were characterized by the presence of two X-chromosomes and males by possessing only one. It soon became clear, however, that the problem was much less simple, for it was found that two X-chromosomes were present in both sexes of many species, some of them closely related to those in which the females have two and the males only

[1] The chromosome which is here for convenience called the X-chromosome has been named by various writers the accessory, the heterotropic, or the hetero-chromosome. The term used in the text has been adopted from American writers to avoid the employment, as far as possible, of unfamiliar technical terms.

Plate XIII

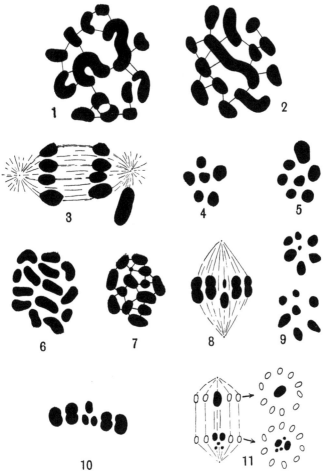

Hetero- and Idio-chromosomes in Insects (Hemiptera). (Reproduced, by permission of the Editor, from *Quart. Journ. Micr. Science*, LIX. (1914), p. 499.) (1–5) *Protenor belfragei*. (1) Female, equatorial plate before reduction, two large hetero-chromosomes. (2) Similar group of the male, one large hetero-chromosome. (3) Spindle of second maturation-division of spermatozoon; large hetero-chromosome below right pole. (4, 5) Two daughter-plates of same division; hetero-chromosome in group 5. (6–9) *Euschistus variolarius*. (6) Female, equatorial plate before reduction; no conspicuously small chromosome. (7) Similar group of the male; one very small idio-chromosome. (8) Spindle of second maturation-division of spermatozoon; large and small idio-chromosomes in centre. (9) Two daughter-plates of same division; small idio-chromosome in upper, large in lower. (10) *Rocconota annulicornis*. Side view of equatorial plate of second maturation-division of spermatozoon, showing large idio-chromosome paired with two small. (11) *Acholla multispinosa*, second maturation-division of spermatozoon, showing large idio-chromosome paired with five small. ((1–10) from Wilson. (11) from Wilson, after Payne.)

v] *Material Basis of Sex-determination* 63

one. It was therefore objected by those who were sceptical about the relation between the X-chromosome and sex that this chromosome could have no connexion with sex-determination, since in some species the male had one less than the female, while in related species the chromosomes of the male and female were alike. An important advance in our knowledge was then made by the American cytologist, E. B. Wilson; he found that in certain species of Hemiptera the male had a large and a small X-chromosome, while the female had two large ones of the same size as the larger of the male (Plate XIII, figs. 6, 7). To avoid confusion, he introduced the term *idio-chromosomes* to designate chromosomes of this kind. He found in a series of related species that one had two equal idio-chromosomes in both male and female, another had in the male one large and one slightly smaller, a third had a large and a very small one, and in others again the small one was completely absent. In all the species the female had two of the same size as the larger of the male. Some American writers have used the expression Y-chromosome for the smaller member of the pair in the male; using this term it may be said that in some species the Y-chromosome is as large as the X-chromosome which behaves as its mate, in other species it is somewhat smaller, in still others it is very small, and finally in many it has disappeared altogether. In all these cases the female has two X-chromosomes, and all eggs therefore have one after the maturation divisions; in the male, since X and Y behave as a pair in the maturation of the spermatozoa, half the spermatozoa will have X and half will have Y. There will thus be two kinds of fertilized eggs, some with constitution XX, others with XY; the former will clearly become females, the latter males.

The condition in which the male has either an unpaired X-chromosome ("accessory" or "heterotropic" chromosome), or in which it has a pair of unequal idio-chromosomes, is widely distributed, not only in some species of most orders of insects (except Lepidoptera), but also in some of the Nematode worms, the Myriapods (Centipedes, etc.), the Spiders, and among vertebrates in the Mammals, including Man. It might naturally be supposed that this condition was universal, and that where no difference could be seen between the chromosomes of the male and female, this was due simply to the fact that the male had two equal idio-chromosomes, one of which is in reality an X-, the other a Y-chromosome. If this hypothesis were accepted, a natural corollary would be that sex is universally determined by the spermatozoa, since in all these cases the eggs are to all appearance alike. It was shown, however, in the discussion of sex-limited inheritance in the last chapter, that there is good evidence that in Lepidoptera and Birds there must be two kinds of eggs, since those which develop into males bear hereditary characters which are lacking in the eggs which become females. Further, it should be noticed that the groups in which sex-limited transmission by the male is known to occur are included among those in which some spermatozoa contain the

Plate XIV

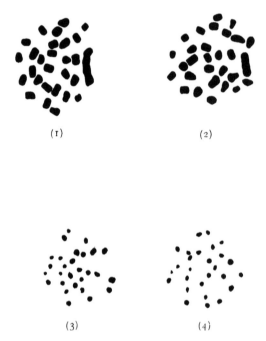

Chromosomes in the maturation of the eggs of Moths. (Reproduced, by permission of the Editor, from *Quart. Journ. Micr. Science*, LIX (1914), p. 506. (1, 2) *Phragmatobia fuliginosa*, after Seiler. (1) Outer group of first polar division, 27 ordinary and one large chromosome (total 28). (2) Inner group of same egg, with 27 ordinary and one large chromosome divided (total 29). (3, 4) *Abraxas grossulariata*, strain which has 55 chromosomes in the female before reduction. (3) Outer group of first polar division, 28 chromosomes. (4) Inner group of same egg, 27 chromosomes. In other eggs the outer group has 27, and the inner group, from which the egg-nucleus is produced, 28.

X-chromosome and some do not; this condition has, in fact, been shown to exist in *Drosophila*, the Cat, and Man, the forms in which sex-limited transmission by the male is best known. This naturally led to the hypothesis that the characters which are sex-limited in transmission are borne by the X-chromosome. There are some grave difficulties in the way of accepting this hypothesis in its simple form, but the behaviour of the X-chromosome corresponds so closely with that of characters which are sex-limited in transmission by the male that it is hard to avoid the conclusion that the two are connected in some way. If this is so, it would be natural that an unpaired or unequally paired chromosome should be looked for in the female of such animals as *Abraxas grossulariata*, in which the sex-limited transmission is by the female instead of by the male. A search for such an unpaired chromosome in *Abraxas* at first led to a negative result; the chromosome number in the female was found to be identical with that of the male. In another moth, *Phragmatobia fuliginosa* (the "Ruby Tiger Moth"), however, J. Seiler found evidence for the existence of a pair of dissimilar idio-chromosomes in the female, while the male had a corresponding pair of equal chromosomes (Plate XIV, figs. 1, 2). Almost simultaneously the writer discovered that although most females of *Abraxas* have chromosome-groups which could not be distinguished from those of the male, some strains have one chromosome missing. All males, and the majority of females, have 56 chromosomes in the nuclei before maturation, giving 28 in the spermatozoa and in the eggs after the maturation divisions. In a special strain, however, which was being used for experiments in heredity, and also in one wild female, it was found that the ovarian cells had nuclei which contained only 55, and that eggs of these females after the first polar division contained in some cases 28 and in others 27 (Plate XIV, figs. 3, 4). It must therefore be concluded that in some females there are a pair of dissimilar chromosomes, corresponding with the so-called X and Y chromosomes (idio-chromosomes) in the male of some other insects, and that in the strain referred to one of these is absent. These observations have thus provided a rather remarkable confirmation of the hypothesis that the presence of an unpaired or unequally paired chromosome is connected with sex-limited transmission, for the hypothesis was founded on cases in which such a chromosome existed in the male, and thence it was inferred that in *Abraxas* and other moths it should theoretically be expected to exist in the female. This naturally led to investigation, and the condition which had been predicted was found.

It is now known, therefore, that in certain species of most orders of insects, and in a number of animals belonging to other groups, the male has one X-chromosome and produces two kinds of spermatozoa, some with and some without X, while the female has two X-chromosomes and produces eggs all of which contain X; in two species of Lepidoptera, on the other hand, the converse condition is found, the male having all its chromosomes paired, while the female has one unpaired or unequally paired. It is obvious from these facts, if the unpaired chromosome in the female of *Abraxas* is of the same character as the unpaired chromosome in the male of other animals, that femaleness cannot universally depend on the presence of an additional X-chromosome. Two possibilities seem open; either in most cases an additional X-chromosome produces a female, but in the Lepidoptera a male, or the unpaired chromosome in the females of Lepidoptera is different in character from the X-chromosome of other forms. Of these possibilities the second seems the more probable.

Before leaving the question of the relations between chromosomes and sex, the problem of the possible connexion between chromosomes and sex-limited inheritance requires some further consideration. It has been explained that all the known cases of sex-limited transmission by the male occur in groups of animals in which the male produces two kinds of spermatozoa, one with and one without an X-chromosome, while the female produces eggs all of which contain the X-chromosome. This condition exactly corresponds with the normal behaviour of the sex-limited characters, which are transmitted by only half the spermatozoa but by all the eggs. On the other hand in *Abraxas* all the spermatozoa are alike in respect of their chromosomes, while there may be two kinds of eggs, one containing a chromosome which is lacking in the other. And correspondingly all the spermatozoa normally bear the sex-limited character, while half the eggs contain it and half are without it. The behaviour of the unpaired chromosome thus exactly corresponds with the normal behaviour of the sex-limited characters in both male and female, and the fact that the existence of an unpaired chromosome was predicted in the female of *Abraxas* on the ground of its sex-limited transmission, still further confirms the belief that the chromosome behaviour is connected with sex-limited inheritance. One serious difficulty remains —that in nearly all cases exceptions occur, though only in a small percentage of the offspring. If the unpaired chromosome were itself the determiner of the sex-limited character, there should be no exceptions at all, for the same chromosome would determine both the sex and the presence of the sex-limited character. Several suggestions have been made to meet this difficulty. One of the most recent is that exceptions in *Drosophila* can be accounted for on the assumption that in the maturation of some eggs the X-chromosomes fail to separate at the polar division (C. B. Bridges). If they are then both extruded with the polar body, the egg-nucleus would be left without an X-chromosome, and such an egg, fertilized by an X-bearing spermatozoon, would be a male, since it contains only one X-chromosome. If now the original female were white-eyed, that is, if it contained two X-chromosomes from each of which the red-eye factor had been lost by mutation, such a female, mated with a red-eyed male, would produce red-eyed male offspring from those eggs from which both X-chromosomes had been extruded, when they were fertilized by X-bearing spermatozoa.

The egg before fertilization, if both X-chromosomes were extruded with the polar nuclei, would contain no X-chromosome; the spermatozoon would introduce one X-chromosome bearing the red-eye factor, and thus the fertilized egg would contain one X-chromosome having the red-eye factor, and would therefore by hypothesis be a red-eyed male. Such a red-eyed male would be an exception to the normal rule that white-eyed females fertilized by red-eyed males give all female offspring red-eyed, all males white-eyed. This explanation is sufficient to account for the particular kind of exception for which it was invented, but there are probably others for which it will not hold. For example, it must be supposed that the chromosome-constitution of the Cat is similar to that of *Drosophila*, since each has the same type of sex-limited inheritance, and since an unpaired chromosome in the male has been discovered in the Cat by von Winiwarter[1]. If the factors for yellow and black were borne only by X-chromosomes, the exceptional tortoiseshell male should then have two X-chromosomes, but if the X-chromosomes are also sex-determiners, such a cat must by hypothesis be a female. In this case, therefore, it seems clear either that the X-chromosomes do not themselves bear the factors for sex-limited characters, or that they are not the cause of sex-determination. Another hypothesis, which in the present state of our knowledge seems more probable, is that the factor which determines the sex-limited character is

[1] By the courtesy of Dr von Winiwarter, of Liége, I am allowed to refer to this discovery before the paper describing it is published.

borne by a chromosome which is usually coupled with the X-chromosome. Such associations between chromosomes have been observed in a number of cases, and it is found sometimes that the X-chromosome is either double, or is usually so closely associated with a second chromosome that the two seem constantly to go together to the same pole of the spindle in the maturation divisions (Plate XIII, figs. 10, 11). If this always happened, any character, the factor for which was borne by the second chromosome, would of course always be carried by those spermatozoa or eggs which bear the sex-determining chromosome; but if the association were occasionally broken, exceptions to the normal sex-limited transmission would appear. A third possibility is that the X-chromosome is not an indivisible unit, and that, occasionally the part of it which bears the factor for the sex-limited character may become separated from the part which is especially connected with sex. This hypothesis differs only slightly from the preceding, but would hold good in such cases as that of *Abraxas*, in which association of distinct chromosomes has not been directly observed.

Finally, it must be made quite clear that the hypothesis that sex is determined by a special sex-factor contained in a particular chromosome is not by any means universally accepted. It seems the simplest hypothesis in view of the obviously close relation existing between certain chromosomes and sex, but it cannot be regarded as proved. Many eminent biologists believe that the chromosome

v] *Material Basis of Sex-determination* 71

behaviour is not the cause, but is, so to speak, a symptom of sex, and that the cause of sex-determination lies deeper. No one believes that the presence of horns is the cause of a Red Deer being a stag instead of a hind; horns are a regular accompaniment of maleness in most deer, but are certainly not its cause. So it may be maintained that an egg or a spermatozoon may have one chromosome more or less because it has in it the power of developing into one or the other sex, rather than that the chromosome is the cause of the sex. This belief has in its favour the fact that in the Aphids and Gallflies the sex of the egg is decided before the maturation divisions which determine the chromosome number, for some females are male-producing and some female-producing. The warning about the use of the word " cause," given in the first chapter, applies especially to such cases as these, for it is possible that the immediate " cause " of the sex is the extrusion or retention of a chromosome, and that this cause is determined by a pre-existing condition; or on the other hand it is at least conceivable that the real cause is this unknown pre-existing condition, and that the sex of the egg would be the same whether the chromosome were extruded or not. In any case, it is highly probable that sex is not determined simply and immediately by the presence or absence of a chromosome. Even if the chromosome behaviour is a necessary link in the chain of causes leading to sex-determination, as seems to the writer probable, it is not the immediate or only cause of the sex. This in one sense is an obvious truism, for the chromosome is part of the cell, and any effects which it may have can only be brought about by its interaction with the other parts of the cell. The most that can possibly be claimed for a chromosome is that its presence in the cells of the body so alters them that the body becomes a female instead of the male, and if the cells should be otherwise altered this effect might not follow. There is evidence, however, which will be considered in a later chapter, that chromosomes, if they are active " causes " at all, do not act in this direct way, but rather are links in a chain of events, of which the determination of sex is only one. At the risk of repetition, an analogy may be given. A locomotive engine has a reversing lever which determines whether the engine shall go forwards or backwards, but the lever is not the cause of the motion of the engine, and by altering other parts of the machinery it could easily be arranged that when the lever was reversed the engine should go forwards. In the same way it is possible that, other things being equal, the presence of a certain chromosome may lead to the development of a particular sex, but it is not impossible that other changes may in exceptional cases so alter the whole mechanism that this effect is not produced.

CHAPTER VI

THE SEX-RATIO

THE general conclusion to be drawn from the facts of sex-limited inheritance and of the relations between chromosomes and sex-determination is that in animals of some groups two kinds of spermatozoa are produced, in others two kinds of eggs, and that, as a result, in each class two kinds of fertilized eggs must exist, one of which grows up into males, the other into females. If this conclusion is correct, and if there is neither differential activity in the two kinds of germ-cells nor differential mortality before birth, it follows that the proportion of the sexes among the young born must be the same as the proportion of the two kinds of germ-cells. When the spermatozoa are of two sorts, both kinds arise by division of one sperm mother-cell and must therefore be formed in equal numbers; and when the eggs are of two kinds, as far as we know it is usually a matter of pure chance whether the sex-chromosome remains in the egg or goes out into the polar body, so that in this case also the two kinds should on the average be equally common. It follows from this that unless there are disturbing causes, the two sexes should be produced, on the average, in equal numbers. It is a matter of common observation that in many animals at least the numbers of the two sexes are about equal, but in general, when the statistics are large enough, it is found that the approximation to equality of males and females, though it may be close, is rarely exact. It is also found that the proportion of the sexes in any species fluctuates more or less considerably according to a variety of conditions, and it is clear that no theory of sex-determination is adequate which does not take these variations from equality of males and females into account.

The relative proportion of males to females in any species is known as the sex-ratio of the species, and it is commonly measured by taking the average number of males per hundred females born. The ratio among adults is of course often widely different from this, for it is probable that in most animals one sex or the other is more liable to death from disease or accident. For 100 females born, it has been found that in Man the ratio of males averages between 103 and 107, in the Rat about 105, in the Horse 98, in the Dog about 118, rising in some breeds to over 140, while some animals have an even greater divergence from equality. The figures sometimes given for invertebrate animals are rarely trustworthy, for they are founded on counts of adults, in which differential mortality may have played a large part. In the case of domestic animals, however, the figures are probably good as far as they go, but even in them there is this element of uncertainty, that there may be differential mortality before birth (abortion). It has been suggested, for example, that the regular excess of male births in Man is due to greater loss of female embryos, but this is not supported by the statistics of still-births, which show a ratio of over 130 males to 100 females. Two possibilities, however, must always be kept in mind ; if it is assumed that the spermatozoon determines sex, the male-producing spermatozoa may perhaps be somewhat more active, or for some other reason, such as slightly smaller size, more successful in entering the egg[1] : if on the other hand the egg has some share in sex-determination, there may be a tendency for the polar divisions to occur rather more often in such a way as to produce eggs of one sex rather than of the other[2]. Because, therefore, the sexes are not exactly equal in number the hypothesis of sex-determination by the germ-cells which unite to form the fertilized egg is not disproved. What is required is to investigate the causes which have been found by observation to influence the sex-ratio, and to relate them, if possible, with chromosome behaviour. Much work of this kind has been done, but it has not yet led to very definite results, and any conclusions drawn from it must for the present be tentative.

The influences which may affect the sex-ratio are of the most varied kinds, some may act on the parents long before the reproductive period, or, in some cases of parthenogenetic reproduction, even on the grandparents ; others seem to take effect on the eggs after

---

[1] Spermatoza of two sizes have been described in insects in which an unpaired chromosome exists.

[2] Some evidence for such a tendency exists in a strain of *Abraxas grossulariata*.

they are discharged from the ovary, possibly by influencing more or less directly the maturation-processes, and others again seem to act on the embryo after fertilization, and to suggest that the hypothesis of irrevocable sex-determination at fertilization is not universally valid. It will only be possible here to give a few illustrative examples, and to discuss the conclusions that may be drawn from the large mass of observations which they represent.

It is an old belief that the sex may be influenced by nutrition, acting either on the embryo itself or on the mother before the eggs are discharged from the ovary. Many of the earlier experiments in feeding part of a batch of larvae well and starving the remainder, are discredited on the ground that the difference in the sex-ratio of the two lots is due to selective mortality, females in general suffering more from starvation than males. More recently, however, it has been clearly shown in Rotifers, Water-fleas and probably Aphids that nutrition of the mother does affect the sex of the offspring.

In the Daphnid (Water-flea) *Simocephalus*, Issakowitsch showed that rich nutrition, and also high temperature, which probably acts indirectly by influencing nutrition, caused the production of parthenogenetic females, while starvation and cold caused the appearance of males. Similar results have been obtained by others, differing chiefly in showing that the changed conditions are only effective at certain times in the life-cycle. Whitney and Shull have shown that pure spring water acts on the parthenogenetic eggs of the Rotifer *Hydatina* during their growth in such a way that they develop into male-producing females, while eggs of females grown in an infusion of horse-manure always give only female-producers.

It has been pointed out, however, that these are really changes, not from production of females to production of males, but from parthenogenesis to sexual reproduction, since a male-producing female, in the Rotifer at least, is a female which produces eggs that can be fertilized, whereas the ordinary parthenogenetic females do not. In all cases in which a long line of parthenogenetic females is caused to produce sexual females and males, the same objection can be made—the stimulus causes a change from parthenogenesis to sexual reproduction, but this is a different thing from changing the proportions of the sexes.

In the case of animals which always reproduce sexually many attempts have been made to influence the sex-ratio by differences of food or other conditions in the parents, or to investigate the effects of such differences when they occur naturally. Heape, for example, has published an account of the sex-ratios produced in two aviaries of Canaries, indicating that differences of temperature and food caused the sex-ratio in one to correspond with 77 ♂ : 100 ♀, in the other 353 ♂ : 100 ♀. The numbers are small, only 200 birds in the first case and 68 in the second, but the results were so consistent over several years that they can hardly be due to chance. Heape regards these results as due, not to a direct action of nutrition on the sex of the egg, but to "forcing" of the reproductive function in one case and retarding it in the other. In one case the birds were caused to breed early without being fed excessively, in the other they were kept back but highly fed; the former gave an excess of males, the latter of females. Heape suggests that male- and female-determining ova are present in the ovary, and that " forcing " in the absence of excess of food encourages the development of the male-determining eggs, while slow development and much food causes the female-determining eggs to develop rather than the male. He accounts in a similar way for the seasonal variation in the sex-ratio of dogs and other animals, including in some cases the human race. In the Greyhound, for example, although there is no perfectly definite breeding season, more young are born in the months March to June than at other times of year. In these months the sex-ratio ranges from 113 to 119 ♂ : 100 ♀. In the winter months, on the other hand, many fewer young are born and the sex-ratio ranges from 128 to as much as 195 ♂ : 100 ♀. Similarly in the semi-civilized population of Cuba, Heape finds that in both white and coloured people there is very considerable correlation between the seasonal fluctuations of the sex-ratio and the seasons of highest birth-rate; in the whites the sex-ratio in the months of highest birth-rate for the three years 1904–6 was 104 ♂ : 100 ♀, in the months of the lowest birth-rate 108 ♂ : 100 ♀; in the coloured people the difference was greater, 99 in the former and 108 in the latter period. These data are based on very large numbers of births, and seem to indicate that where there is a faintly marked breeding season, the proportion of males is significantly lower among the offspring born in the breeding season than among those born at other times of year. The breeding season depends on the physiological activity of the organism, and Heape's contention is that when metabolic activity is great, female ova tend to develop rather than male, while when metabolism is low, male ova develop more easily than female.

Heape regards the female as having the preponderating influence in all cases, and other observations and experiments, to be mentioned shortly, may be used in support of this opinion, but there is some evidence that the male parent is also not without influence on the sex-ratio. It is a wide-spread belief among breeders of some animals, for example Cats, that the average proportion of the sexes in the litter depends on the age of the father, that young males have more female and old males more male offspring. Several writers have brought forward evidence pointing in the same direction in the case of Man, but others, working on different statistical material, have found no support for the idea that the ages of the parents have any influence. Others, again, regard the age of the mother as of more importance in this connexion than that of the father; for example Ewart has noted that in the birth-statistics of Middlesbrough the ratio of males to females rises regularly from 70 ♂ : 100 ♀ for mothers less than 20 years of age, to 116 ♂ : 100 ♀ for mothers over 34. His data are not extensive—less than 2000 births in all—

but the regularity in the rise of the ratio of males to females with increasing ages of the mothers is remarkable if it is due merely to chance, and similar conclusions have been drawn by a number of other investigators, both from human statistics and from those of horses and other domestic animals. It is almost certain, however, that no universally valid statement can be made about the relation between the age of the parents and the proportion of the sexes in the offspring, for not only is there disagreement among investigators of one and the same species, but also conclusions which seem trustworthy in the case of one species are certainly inapplicable to others.

One of the chief difficulties of obtaining any reliable results from statistics of sex-proportions is the impossibility of making the necessary corrections for possible causes of error. An interesting example of this is provided by Punnett's attempt to discover whether nutrition affected the sex-ratio in Man. He divided the boroughs of London into three groups, A with less than 15 indoor servants per 100 families, B with between 15 and 30 servants per 100, and C with over 30 servants per 100 families. Reasons are given for believing that the number of domestic servants is a fair index of wealth and of nutrition. The ratio of male to female infants for each class was then calculated from the data given in the 1901 census returns, and when the numbers had been corrected for the greater mortality of males, they gave in the poorest class, A, 101 ♂ : 100 ♀ , in class B, 102·2 : 100, in class C, 103·7 : 100, and finally the ratio calculated from 5225 births recorded in Burke's *Peerage* was 107·6 ♂ : 100 ♀. By themselves these results might suggest that poor nutrition increases the proportion of females, exactly the opposite result to that obtained by many workers on animals. Punnett points out that there are several considerations which render such a conclusion untenable. Firstly, the infant mortality is higher in the poorer districts, and probably falls most heavily on the males, so that the census records give too low a ratio of males in class A. Secondly, he gives evidence, which is in accord with that obtained by others, that the first and second children in a family show a greater excess of males than the later children, and since there is a greater tendency towards limitation of the size of the family among the more well-to-do, this factor tends to raise the ratio of males in the richer classes. Finally, Punnett's data agree with those already mentioned in showing a lower ratio of males among the children of young mothers, and since the richer classes in general marry later, this also would tend to raise their proportion of males. Punnett therefore concludes that the apparent deficiency of males in the poorest part of the population is not a consequence of defective nutrition, but is due to the combined effects of the causes mentioned[1].

[1] A curious instance of the way in which abnormal sex-ratios may give rise to misleading inferences has been given by Morgan. It was found that a certain strain of *Drosophila* had a ratio of about 1 ♂ : 2 ♀, and that the tendency to produce offspring in this proportion was inherited through the female. It was consequently inferred by the writer and others that sex could not be determined in *Drosophila* solely by the spermatozoon, as had been assumed from

It would seem from cases like those described that in consequence of the many factors which influence the sex-ratio, little positive information can be obtained from it about the causes which determine sex. There are nevertheless some definite experimental results which are of considerable importance in this connexion, since they show that in some cases at least the proportion of the sexes can be very largely altered by artificial means. Of these the work of R. Hertwig on Frogs may be taken as an example. Hertwig investigated especially the effects of early and late fertilization on the eggs of the frog. He allowed a frog to lay some eggs normally; it was then separated from the male and kept for 64 hours with the remaining eggs in its oviducts, and these were then fertilized by the same male. The first eggs gave a normal sex-ratio, those fertilized after 64 hours gave a ratio of almost 700 ♂ : 100 ♀. Kuschakewitsch, repeating Hertwig's experiment with a longer interval before fertilization (89 hours), succeeded in getting a culture consisting entirely of males. It has been objected that this excess of males is due to the death of the female embryos, but since in some experiments the mortality was quite

the chromosome behaviour. Morgan then showed that the result was probably due to a sex-limited "lethal factor," which causes the early death of all individuals in which its effect is not counterbalanced by a normal factor received from the other parent. Since these factors are sex-limited, only one or the other, but not both, can exist in the male; therefore all males which contain the lethal factor fail to develop. A female which is heterozygous for the factor transmits it to half her offspring of each sex, and thus half the males disappear, and a ratio of 2 ♀ : 1 ♂ results. Morgan's evidence in favour of this explanation is almost conclusive.

small, this cannot account for the result. It was also proved that it could not be caused by over-ripeness of the spermatozoa. Hertwig makes the not improbable suggestion that the effect is caused by a change in the character of the maturation divisions of the egg-nucleus. The polar bodies are given off after the egg is laid, and he suggests that over-ripe eggs tend more or less constantly to extrude the female-determining " factor " (an X-chromosome or its equivalent), while in normal eggs it is a matter of chance whether it goes into the polar nucleus or remains in the egg-nucleus. A suggestion of the possible nature of over-ripeness has been given by Miss H. D. King, who finds that by treating the eggs of Toads in various ways which reduce their water-content, the proportion of females is raised very considerably, while by treating them with very dilute acid, which causes them to absorb water, the proportion of males is correspondingly increased. It may perhaps be concluded that increase of water-content tends to cause the female-determining factor to be extruded with the polar nuclei, while decrease of water causes it to be retained in the egg-nucleus.

Similar results have been obtained in other animals, for example Pearl has given evidence that when fertilization takes place in cattle some considerable time after the ovum is discharged from the ovary, the ratio of males is greatly increased. The same thing is said to be true of Dogs and possibly Horses. In this case it should be noted that in the few Mammals which have been examined, the female has two X-chromosomes and the male one, so that all eggs must contain an X-chromosome, and to such cases Hertwig's hypothesis of its expulsion by the polar body in the over-ripe egg is clearly inapplicable. If all Mammals are alike in this respect, Hertwig's explanation will thus not account for the effects of over-ripeness of the ova in Cattle, unless it be assumed that the two X-chromosomes bear different sex-factors, and that over-ripeness of the ovum tends always to cause the retention of one rather than of the other. Such an admission would practically involve the abandonment of the hypothesis that the presence or absence of two X-chromosomes is the cause of sex-determination.

Before leaving the subject, there are other experiments by Hertwig which must be referred to. Having shown that in the Frog over-ripeness of the eggs before fertilization caused great excess of males, he tried to discover whether the male parent also had any effect on the sex-ratio. He found, by fertilizing eggs with spermatozoa from males at very different stages of their breeding season, that the age of the spermatozoa had no effect. But at the same time he proved that different males give conspicuously different sex-ratios among their offspring. For example, eggs of a single female fertilized by sperm of one male gave 52 males and 50 females, while other eggs of the same female fertilized by another male gave 3 males and 64 females. Further, he found that in frogs from certain localities the sex is not visibly differentiated until at quite a late stage in the growth of the young frog. In most young frogs the testes or ovaries are clearly recognisable at an early stage, but in these particular localities all the young frogs were " indifferent "—it was impossible to tell by dissection whether they were males or females. Since the sex-ratio of adult frogs in these localities is about normal, it is clear that the "indifferent" young must develop into males in some cases and females in others. Experiment showed that this " indifferent " condition could be determined by the male parent; for example, eggs of one female fertilized by one male gave 140 males and 142 females, while other eggs of the same female fertilized by another male gave 109 indifferent young. Other cases gave a mixture of males or of both sexes and indifferent larvae, and it was also proved that the indifferent condition may be caused by the eggs of females belonging to the race in which indifference occurs. Since in several cultures the indifferent individuals seem to take the place of females (for example, one female gave 52 ♂ : 50 ♀ with sperm from one male, 54 ♂ : 69 indifferent with a second), Hertwig regards the indifferent genital organ as a rudimentary ovary, with its female characters so diminished that it may develop into a testis. He suggests that the X-chromosomes of the race with indifferent young are weaker in their effect than in normal frogs, and that sometimes they are not effective at all, and the indifferent young become males. Pearl has similarly suggested that the effects of staleness of eggs is artificially to weaken the action of the sex-chromosome, and so accounts for the preponderance of males from over-ripe eggs.

Finally, experiments by Hertwig on the effects of

conditions acting after fertilization give further indications pointing in the same direction. He divided the fertilized eggs of one female into two parts, and kept one lot first at 15° C. (59° F.), later at 16°–18° C. (61°–64° F.), the other lot at 30° C. (86° F.). Those kept in warm conditions metamorphosed in about a month; those in the cold at varying times between six and ten months. The warm culture gave 344 males, 319 females, the cold 260 males and 85 females. It is true that there was heavy mortality in the cold culture, but Hertwig regards this as insufficient to account for the difference.

One other quite distinct cause of the production of abnormal sex-ratio must finally be mentioned, to a special case of which further reference will be made in a later chapter. It has frequently been noticed that hybrids between distinct species, or even between varieties of one species, show conspicuous deviations from equality of the number of males and females. In the case of moths, some crosses produce only or almost exclusively males, although the converse cross between the same two species may yield an excess of females. Many less extreme instances have been described in Vertebrate hybrids; as examples may be chosen Mrs Haig Thomas's experiments with Pheasants, and a recent paper by J. C. Phillips on Ducks. In various crosses between Silver and Swinhoe Pheasants, also between Golden and Amherst, Formosan and *versicolor*, and between Reeve's Pheasant and Formosan and *versicolor*, Mrs Haig Thomas reared a total of 228 males and 135 females, a ratio of 168 ♂ : 100 ♀. It is true that about two thirds of the eggs laid did not develop at all, and that there may have been an excessive mortality of female embryos in the early stages, but it is certain that in these hybrids there is a great excess of males among the embryos which are capable of surviving to the point of hatching. Somewhat analogous results are recorded by Phillips in the case of crosses between the male of a large domestic breed of Duck (the Rouen) with Wild Duck (Mallard) females. He obtained 46 ♂ : 24 ♀, a ratio of 192 ♂ : 100 ♀. In this case he suggests that the result may be connected with the great difference of size between the parents, since when the crossed ducks were mated together their offspring showed a ratio nearly approaching equality. Considerable excess of males has been observed in crosses between species or varieties of Rats and Mice and, lately, there is some evidence that in Man marriages between distinct races may produce offspring among which the proportion of the sexes differs from that of either pure race. Pearl has collected data from the Argentine Republic which indicate that the mating between an Italian father and an Argentine mother gives a ratio of males (105·7 : 100) higher than that of either pure Italian (100·7) or pure Argentine (103·2), and similar, though less striking, differences were found among the children of Spanish and Argentine parents.

It is noteworthy that in all these cases, and apparently generally when abnormal sex-proportions appear as the result of hybridization, the excess is of males rather than of females. This may be compared with the excess of males produced from eggs of which the fertilization is delayed, though our ignorance of the causes involved in each case is still too great to make it possible to decide whether the apparent similarity between the two cases is more than accidental.

It will be most convenient to defer to a later stage any critical discussion of the facts outlined in this chapter, and here merely to give a short summary of them. In general, an examination of sex-ratios as they occur naturally, and of the effects of external conditions in altering them, shows that the ratio at birth of males to females varies from a variety of causes: some of these are unconnected with the real problem of sex-determination, since they are concerned with differential mortality, but others are of more fundamental importance. In most cases the external influences which affect the sex-ratio seem to act before fertilization, in many cases perhaps by altering the proportion of male-determining and female-determining germ-cells. Such environmental effects are not inconsistent with the hypothesis that sex is determined by specific " factors " borne by the ova or spermatozoa. On the other hand, some examples are described in which the effects of environment seem to act in the egg after fertilization, or upon the egg in animals in which it has been maintained that the spermatozoon is of primary importance in determining sex. Further, Hertwig's experiments with Frogs show that both the eggs and the spermatozoa have the power of influencing the sex-ratio in the same species. All these latter cases are incompatible with the belief that sex is always exclusively determined by the spermatozoon in some species, and by the egg in others, and that the sex is irrevocably fixed from the moment of fertilization. On the whole, therefore, the study of the sex-ratio, while not leading to any positive conclusion with regard to sex-determination, makes it necessary to reconsider the simple hypothesis of final determination by one or other of two kinds of germ-cells, to which the facts of sex-limited inheritance and of chromosome behaviour seem naturally to lead.

CHAPTER VII

SECONDARY SEXUAL CHARACTERS

THE preceding chapters have been concerned almost entirely with the various possible causes which may determine whether a fertilized egg shall develop into a male or a female, and except in connexion with sex-limited inheritance nothing has been said about the development of the bodily characters which are commonly associated with one or the other sex. In some respects this is an entirely different question, which might seem to have little connexion with sex-determination, but recent work has shown that the two subjects are so intimately related that it is impossible adequately to consider one without the other. It was pointed out in the first chapter that although the sexes of animals may best be defined according to whether they produce egg-cells or spermatozoa, there are more or less pronounced physiological differences between them, and that it is not certain whether these differences are a consequence or a cause of egg-production in the one sex and of sperm-production in the other. In some cases the physiological differences are of such a kind that they can be detected by the methods of physiological chemistry; the blood and other tissues of the body can sometimes be shown to be of different constitution in the two sexes of the same species. More frequently the differences, although not directly demonstrable, are indicated by bodily features which are characteristic of the two sexes, and when these features are not immediately concerned with reproduction, they are spoken of as "secondary sexual characters." In the majority of the higher animals, especially in many Insects, Birds and Mammals, adults are recognisable at a glance as either males or females by differences of size, proportion or colouring, and most frequently these distinctive sexual characters have no immediate or necessary connexion with reproduction, although they may sometimes be concerned with it indirectly, for example by attracting mates or by being used to vanquish rivals. Conspicuous differences between the sexes are not so common in the less highly organised animals, although they occur here and there in nearly all the larger groups, and, in general, sexual distinctions of some kind, apart from the reproductive organs, are the rule rather than the exception.

Apart from secondary sexual characters in the more restricted sense of the words, very many animals have sexual differences which are directly connected with reproduction, but are not immediately related to the production of eggs or spermatozoa. Female mammals have milk-glands for the nourishment of the young; many female insects have a boring apparatus (ovipositor) by which the eggs are inserted into the food-plant; the males of several fishes and frogs have special sacs for carrying the eggs, which are placed in them as soon as they are laid by the female, and remain there till they hatch. Such examples could be multiplied indefinitely, and the impossibility of sharply distinguishing these accessory reproductive structures from secondary sexual characters is illustrated by the sting of the bee or wasp, which is a typical secondary sexual character and yet is a modified ovipositor.

We find these sexual distinctions almost universally distributed, and ranging from minute differences, to be detected only with difficulty, on the one hand, to extreme and sometimes almost ridiculous sexual dimorphism on the other, as in the Peacock and the worm *Bonellia* (Pl. I), and it is hardly possible to doubt that they depend on more or less considerable differences of physiological constitution, whether these can be directly observed by our rather crude methods or not. The problem then arises, what is the cause of these differences in the physiology of the sexes? Are they a consequence of the presence of an ovary in one individual and a testis in another, or are they due to inherited " factors," comparable with those which determine the presence of ordinary Mendelian characters such as the comb-form in fowls or the coat colour of rabbits? If the latter alternative is the true one, how is it that they are so intimately correlated with sex? Or, again, is it possible that both these suggestions are mistaken, and that not only sex, as defined by the presence of ovary or testis, but also other distinctively sexual characters are the consequence of fundamental physiological differences, which may or may not be demonstrable apart from their effects on bodily form?

It is clear that until some answer can be given to these questions, or some means be found of choosing between the mutually exclusive possibilities, no theory of sex-determination can be regarded as complete. Biology is still very far from giving an answer which is applicable to all cases, and it is becoming clearer almost every day that solutions which seem satisfactory in one instance are quite inapplicable to others. It is difficult to believe, however, that the problems of sex are so distinct in different animals that no common basis exists, and therefore if a solution which seems adequate for one species is found not to apply to another, the probability is that it is really only a partial solution, and that the real answer to the problem must be sought at a deeper level. The final explanation has not yet been found; the purpose of the present chapter is to indicate the more important results to which the search has led, and the direction to which they seem to point.

The first question which seems to require an answer is whether the secondary sexual characters are directly dependent on the presence of the ovary or testis. Generally they appear, in their full development at least, only at sexual maturity, and in many animals which have a restricted breeding season they are shown to their fullest extent only when the reproductive functions are most active. From these facts alone it might be inferred that they are to some extent at least dependent on the functional activity of the reproductive organs, and in a number of Vertebrates this conclusion is definitely confirmed by experiment. Castration of the male in mammals before sexual maturity prevents the full development of secondary sexual characters; a male deer if castrated when young grows no antlers, and if the operation is performed after they have attained their full size, when they have been shed they do not grow again, or at most develop only to a small extent, in the following seasons. In sheep some breeds have horns in the male and none in the female; other breeds are horned in both sexes, but the rams have larger horns than the ewes. Marshall has shown that if a ram of a breed in which only the male is horned is castrated before the horns begin to grow, no horns, or only small "scurs" are produced; when it is castrated when the horns are still small, they stop growing from the time of the operation. A castrated male of a breed which is horned in both sexes develops horns like those of the female. The horns of sheep will be referred to again in connexion with the inheritance of secondary sexual characters, but the facts given, together with many other experiments of a similar kind, show clearly that in mammals the secondary sexual characters of the male are very largely dependent on the presence of a functional testis. The removal of the testes before they have become fully functional prevents the development of the male secondary sexual characters, but generally does not cause the appearance of those normally found in the female.

Removal of the ovary in mammals has less effect

than castration of the male; it prevents the development of the typical female sexual characters, so that the animal remains somewhat juvenile in appearance, but does not, usually at least, lead to the development of any of the distinctively male features.

It has been known for some time that the effect of the testis or ovary in stimulating the development of the distinctively sexual characters is probably due, not to the existence of functional spermatozoa or ova, but to the part of the organ known as the "interstitial tissue," and it has been supposed that the effect is brought about by the secretion from this tissue of special substances, called "hormones," which circulate in the blood and stimulate the parts concerned to produce these special characters. The question of "hormones" will be considered more fully later, but the importance of the interstitial tissue, as contrasted with the actual germ-cells, in this connexion may be illustrated by a reference to a recent paper by E. Steinach. Steinach removed the ovaries and testes from a number of very young rats and guinea-pigs; some were merely castrated, into others he grafted testes or ovaries from other individuals of the same sex, and in others again he grafted an ovary into a male and a testis into a female. The animals which were simply castrated developed into adults that were normal except that the distinctive secondary sexual characters were reduced or absent. The testes which were grafted into other males never developed spermatozoa, but the interstitial tissue was developed to an abnormal extent, and correspondingly these males had their distinctively male characters—size, proportions, fighting and sexual instincts—excessively pronounced. Males into which an ovary was grafted were " feminized," and grew up into animals with the appearance and habits of females, and, most remarkable result of all, with milk-glands so developed that when given young ones they were able to suckle them, which they are said to have done with evident satisfaction. The successful transplantation of a testis into a female was very difficult, but when the operation was successful the animal was " masculized "; it grew large and powerful, fiercely attacked males, and showed the sexual instincts of the male towards females.

Before considering the conclusions which may be drawn from such experiments as these it will be best to give some account of the results obtained from similar experiments on Birds. The " caponizing " or castration of the male fowl is a familiar operation, and the results, which are generally known, have been confirmed by careful experiments by several investigators. In the fowl castration of the male prevents the appearance of some of the characteristic secondary sexual characters. The comb of the capon is smaller, the spurs poorly developed, if present at all, and the male instincts are suppressed. The plumage, however, is but slightly affected, so that the capon resembles a cock much more nearly than a hen. Removal of the ovary, on the other hand, causes the female to assume to a considerable extent the male characters; the comb becomes like that of the cock, and much of the plumage is male in character, interspersed with feathers like those of a normal hen. Similar results were obtained by Goodale with Ducks; the castrated male retained the typical drake plumage, but did not assume the peculiar summer plumage which somewhat resembles the usual colour of the female[1]. Removal of the ovary from the female, however, caused it to assume very gradually the typical plumage of the drake. In both birds, therefore, castration of the male had a relatively small effect, but removal of the ovary from the female caused the assumption, more or less completely, of the male plumage. This is in accord with the well-known fact that old female birds, when past the reproductive period, tend to assume characters typical of the male.

Taking these facts from mammals and birds together, it might seem that the secondary sexual characters were produced as a direct consequence of the activity of the testis or ovary, the testis having apparently the more important effect in mammals, the ovary in birds. Further, since it is probable that it is not the germ-cells themselves that are of importance in this connexion, but rather the interstitial tissue[2], which is supposed to secrete "hormones," it might be concluded that specific hormones are produced by the testis and the ovary, and that the male hormone is the sole cause of the development

---

[1] In similar experiments by Poll, the castrated drakes moulted normally.

[2] Miss Boring, on the contrary, maintains that there is no interstitial tissue in the bird's testis.

of the male secondary sexual characters, the female hormone of those of the female. Several facts, however, show that this explanation is quite inadequate. In Insects, as will be described almost immediately, the secondary sexual characters are independent of the presence of the testis or ovary, and this alone would be sufficient to throw some doubt on the explanation suggested by the vertebrates. There are also facts derived from the vertebrates themselves which make the hypothesis untenable in its simple form. One of the most convincing is the existence, in very rare cases, of " gynandromorphs "—individuals which have the characters of one sex on the right side of the body and of the other sex on the left. Such gynandromorphs are not uncommon in insects but are extremely rare in vertebrates; more than one example, however, has been observed in birds. Poll has described a Bullfinch which had the male and female plumage sharply separated on the two sides of the body; the right side of the breast was red like a normal male, the left side grey like a female. On the right side the bird had a testis, on the left an ovary (Frontispiece). A similar case is recorded by Weber in the Chaffinch. A Pheasant showing somewhat similar gynandromorphism has been fully described by Bond. In this bird the male and female plumages were less sharply separated, though on the whole the left side was preponderatingly male, the right female. There was a single reproductive organ in the left side in the usual position of the ovary, and this contained both ovarian and testicular tissue.

No trace of reproductive organ was found on the right side. As Dr Bond has pointed out, if the secondary sexual characters were determined simply by the presence of male and female " hormones," a condition such as occurred in these birds would be impossible, for both kinds of hormones would circulate in all parts of the body, and the sharp line of separation between male and female characters along the greater part of its length would be entirely inexplicable.

It follows from these facts that the secondary sexual characters cannot arise simply from the action of hormones; they must be due to differences in the tissues of the body, and the activity of the ovary or testis must be regarded rather as a stimulus to their development than as their source of origin. The case of the gynandromorph bird shows that the tissues of the male must differ in some way from those of the female, and if the hormone hypothesis is correct, the male tissues must respond to the hormone produced from the testis, the female tissues to that from the ovary, to give rise to some at least of the typical sexual characters. The extent to which the tissues of the two sexes differ must vary in different animals and in respect of different characters; in birds the facts just described show that the difference must be considerable, while in mammals Steinach found that each sex could be made to assume to a very great extent the characters of the other, by cross-grafting of ovaries and testes. Further evidence pointing in the same direction is given by the experiments on Insects and Crustacea described below and by work on the hereditary transmission of secondary sexual characters, an account of which is given later.

Our knowledge of the physiology of the secondary sexual characters of Insects has been greatly increased by the recent work of Meisenheimer and Kopeč, supplemented by an important paper by Steche. Meisenheimer removed testes and ovaries of young larvae of the moth *Lymantria dispar* ("Gipsy Moth") (Pl. XVIII, p. 115), a species in which the male is dark in colour and has much-feathered antennae, the female very pale and with antennae only slightly feathered. In none of the castrated males was there any alteration of the normal male characters; some of the castrated females were slightly darker than the normal, but in general the female characters were typically developed. Meisenheimer, and afterwards Kopeč, also found it possible to graft ovaries into males and testes into females; the transplanted genital organs persisted and grew and sometimes became connected with the genital ducts, the testis attaching itself to the oviduct and the ovary to the sperm-duct. Even in these cases the moths when they hatched had the characters of the sex to which the larvae belonged before the operation; their sexual instincts were also unaltered, so that a male containing ovaries instead of testes would pair readily with a normal female. To determine whether the secondary sexual characters are already fixed at a stage before the operation, the wing-rudiment was removed from one side of some of these grafted larvae, and the regenerated wing had the same characters as the untouched one; it seems clear, therefore, that in moths the sexual characters are independent of the presence of ovaries or testes.

An interesting observation for comparison with the results of these experiments has been made by Steche. He finds that the blood of the male and female of moths of the same species is very different; in some cases the difference is visible, the male blood being yellow and that of the female green, owing to the presence of the green chlorophyll from the plants on which the larvae feeds. When tested by physiological methods (precipitin test) he finds nearly as much difference between the blood of the two sexes of one species as between bloods of the same sex of distinct species. Steche concludes that there is a fundamental physiological difference between the tissues of male and female in moths, and hence it is impossible to transform the male characters into those of the female by grafting an ovary in place of a testis[1].

Somewhat similar results have been arrived at by Geoffrey Smith in his series of studies on Crabs, of the genus *Inachus* infested with the parasite *Sacculina*.

[1] In this connexion reference may be made to the experiments of Standfuss. He finds that in certain butterflies the effect of high temperature acting on the pupa is to cause the female to assume some of the male secondary sexual characters, while low temperature may cause the male to resemble the female in colour. Since in these cases the male is normally more brilliant than the female, it seems probable that higher temperature increases, lower temperature decreases, the brilliancy, in the one case causing the female metabolism to approximate to that of the normal male, in the other depressing the metabolism of the male towards that of the normal female.

Sacculina is an internal parasite, which is shown by its free-swimming larva to be related to the Barnacles (Cirripede Crustacea); in its adult condition part of its body projects to the exterior under the abdomen of the crab in which it lives, while root-like processes which absorb the juices of its host ramify to all parts of the crab's body (Pl. XV, fig. 1). The roots avoid the more vital organs, and absorb nourishment chiefly from the blood. *Sacculina* may occur in crabs of either sex; in both sexes its presence causes atrophy of the genital organs, so that infected crabs are sterile; it also stops growth so that they do not moult. In the female crab comparatively little effect is produced on the secondary sexual characters; they are somewhat reduced, but the abdomen remains relatively wide and its appendages are still present and have the hairs to which in a normal female the eggs are attached. The male crab, on the other hand, when attacked by *Sacculina*, undergoes profound changes (Pl. XV). Its abdomen becomes wide like that of the female, and the posterior abdominal

DESCRIPTION OF PLATE XV.

Sacculina and its effects on its host. (1) Diagram showing a *Sacculina* attached to the under side of the abdomen of a Shore Crab, with its roots ramifying in the body and limbs of its host. (After Delage.) (2–9) Spider Crabs, *Inachus dorsettensis*. (2) Normal male. (3) Under side of abdomen of normal male, showing copulatory styles. (4) Normal female. (5) Under side of abdomen of normal female, showing feathered appendages for carrying the eggs. (6) Sacculinized male. (7, 8) Under side of abdomen of two sacculinized males, showing reduction of copulatory styles and development of feathered appendages. (9) Under side of abdomen of sacculinized female, showing reduction of appendages. (2–9, by permission from Geoffrey Smith.)

Plate XV

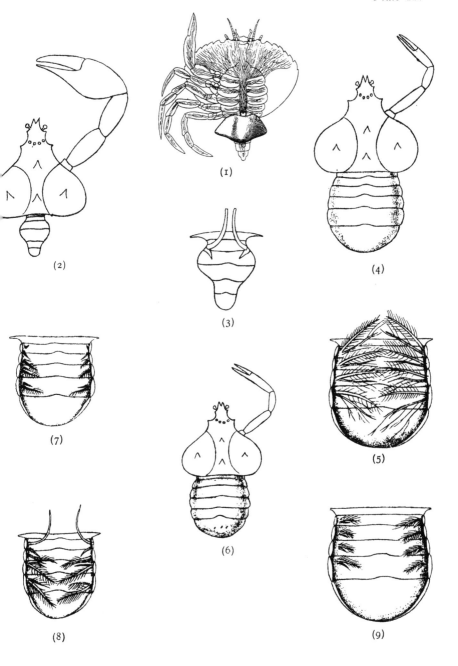

appendages, which in the normal male are absent, develop and resemble more or less those of a normal female. The sacculinized male, in fact, is "feminized" as regards its secondary sexual characters, and in one case at least, in which the parasite had died and the crab's genital organs had begun to develop again, rudimentary eggs were found in the testis of a recovering male.

It was formerly supposed that "parasitic castration" by *Sacculina* was of the same nature as experimental castration such as has been described above in other animals, and that the assumption by the male of the typically female characters was comparable with the assumption of the male characters by a female bird from which the ovary had been removed. Geoffrey Smith's later work shows that the facts are not so simple. In the first place, the parasite does not destroy the genital organs directly; they degenerate under the influence of its action, but it does not actually penetrate into them. Secondly, Geoffrey Smith finds that the blood of the female crab differs in chemical constitution from that of the male; it contains fatty substances which are absorbed by the ovaries and used in the production of the yolk of the eggs. These fatty substances form an important part of the food of the *Sacculina*, and since the normal physiology of the female is such as to be able to produce them in large quantities as fast as they are used up in forming the yolk of the eggs, the action of the parasite, in this respect at least, merely imitates what normally happens in the ripe female, and stimulates the production of more of these substances. As fast as they are produced the parasite absorbs them, so that the eggs are unable to grow and the ovary degenerates. In the male the condition is different; the substances on which the parasite chiefly feeds are present in quite small quantity in the blood; some amount of them is, however, necessary to the life of the crab, and since they are absorbed by the *Sacculina* much more quickly than the crab normally produces them, its whole physiology has to be altered in order to meet the demands of the parasite. As in the normal female, its nutritive processes ("metabolism") become largely devoted to the production of these yolk-forming substances, and although as fast as they are fully formed they are abstracted by the *Sacculina*, the intermediate substances from which they are derived must be supposed to be present in the blood of the sacculinized male just as they are in the normal female.

These facts throw important light in the development of the female sexual characters, and, after recovery, of eggs, in the parasitized male. Taken together with Steche's observation on the different metabolism of the male and female in moths, they suggest that the development of the female characters in the male of a crab infested with *Sacculina* is not the consequence of the atrophy of the testis, but of the change in nutritive physiology, induced by the parasite, from the male type towards the female type. The parasite compels the male crab to produce yolk-forming substance in its blood, and in so doing causes its whole physiology to approximate to that of the female. Concurrently with this the secondary sexual characters come to resemble those of the female, and when the abstraction of the yolk-substances ceases by the death of the *Sacculina*, the testis may even develop eggs. The natural conclusion from these facts is that the difference in physiology is the cause of the change in sexual characters, and that the crab comes to resemble a female because the physiology of its body-tissues has been changed from the male to the female type.

The idea outlined in the last paragraphs has two important theoretical consequences. In the first place, it suggests that it is not impossible that the sex of an embryo may be modified after fertilization if the physiological condition on which sex depends can be changed; to this subject further reference will be made in a later chapter. Secondly, it has led Geoffrey Smith to put forward a hypothesis of the mode of action of the genital organs in influencing secondary sexual characters which differs considerably from the hormone hypothesis outlined earlier in this chapter. If the hormone hypothesis is correct, it should be possible to cause the appearance of secondary sexual characters in castrated animals by injecting extract of the testis or ovary. The evidence with regard to the effects of such injection is conflicting, and Geoffrey Smith believes that most of the positive results which have been obtained are due to misinterpretation or to error. He suggests as an alternative hypothesis that the testis or ovary acts in the same manner on the rest of the organism as does the *Sacculina* on the crab; in his view the reproductive organ removes certain substances from the blood and thus stimulates the production of these substances in excess, and it is these substances, produced in consequence of the action of the reproductive organ, but not from it, that constitute the so-called hormones which bring into existence the secondary sexual characters.

CHAPTER VIII

THE HEREDITARY TRANSMISSION OF SECONDARY
SEXUAL CHARACTERS

In introducing the subject of sex-limited inheritance in a previous chapter, it was mentioned that objection had been taken to the word "sex-limited" on the ground that it might lead to confusion with the inheritance of secondary sexual characters, which are of course limited to one sex. When it is understood that the word "sex-limited" refers simply to the transmission, indicating that the character in question is transmitted, by one sex, only to offspring of the other sex, there is no danger of confusion. It is quite possible, however, that there may be a real relationship between sex-limited transmission and the inheritance of secondary sexual characters, and it is to this latter subject that the present chapter is devoted.

If a character were always found in one sex and never in the other, it would be possible to regard its inheritance as sex-limited in a somewhat different sense from that used above. Sex-limited transmission is defined as the transmission by one sex of a character only (or almost exclusively) to individuals of the other sex, as the *grossulariata* character in the

Currant Moth is transmitted by the female only to her male offspring. If, however, a character were transmitted by one sex only to individuals of the same sex, it might never appear in the other sex at all and would be a secondary sexual character. If, for example, a peacock transmitted his gorgeous plumage only to his male chicks, and if the " factor " for the male plumage were never present in the female, it would always be present in males and never in females and an explanation would be provided of the inheritance of secondary sexual characters. Experiment shows that the facts are rarely, if ever, as simple as this. It was shown in the last chapter that female birds may assume more or less completely the male plumage when their ovaries are removed or degenerated, and this in itself would suggest that the hypothesis would not hold. It is true that castration of the male bird does not usually cause it to assume the plumage of the female, and from this it might be maintained that the female transmits the factor for her plumage only to her daughters, and that it is the male which lacks the factor for the female plumage, not the female which lacks that of the male. Experiments like those of Steinach on rats and guinea-pigs, however, show that this explanation is not universally applicable, and it is still more conclusively disproved by the results of crossing different races or species.

It has long been known that in Fowls and other birds characters peculiar to the male of one breed may be transmitted by the female when crossed with another breed, and it has recently been proved by

Mrs Haig Thomas not only that this is so in various species of Pheasants, but also that the male can transmit the plumage peculiar to the female. In one experiment Mrs Haig Thomas mated a female Formosan Pheasant (*P. formosus*) with a male of the Japanese species (*P. versicolor*). The first-cross offspring already showed that each sex can transmit the secondary sexual characters of the other, for the males had some of the characters of the male *formosus*, the females some of those of the female *versicolor*. The transference of the female characters by the male was still more clearly proved in the second generation; one of the hybrid females was mated back with the *versicolor* male, and all the female young produced (five) had all the typical characters of pure *versicolor* females. In this case there was no *versicolor* female in the ancestry, the crosses being made thus:

formosus ♀ × *versicolor* ♂
 |
hybrid ♀ × *versicolor* ♂
 |
versicolor ♀ (5)   hybrid ♂ (2)

The female offspring of the second cross were nevertheless pure *versicolor* in their secondary sexual characters. Their two brothers still showed considerable traces of their hybrid ancestry. Similar results were obtained by the same investigator with crosses between the Swinhoe and Silver Pheasants, in which it was shown that the Swinhoe male transmits the plumage-characters of the Swinhoe female, thus:

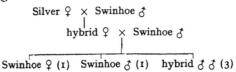

Silver ♀ × Swinhoe ♂
 |
hybrid ♀ × Swinhoe ♂
 |
Swinhoe ♀ (1)   Swinhoe ♂ (1)   hybrid ♂ ♂ (3)

Many similar cases are known. In Moths, *Biston hirtaria* (the "Brindled Beauty") has full-sized wings in both sexes; a related moth, *Nyssia zonaria* (the "Belted Beauty") has a wingless female. Yet *hirtaria* females mated with *zonaria* males produce female offspring with much-reduced, flightless wings, although both the actual parents are fully winged (Pl. XVI). In a plant-bug, *Euschistus variolarius*, the Misses Foot and Strobell have shown that a secondary sexual character of the male, a peculiar black spot, is transmitted by the female. They crossed the female *E. variolarius* with the male of another species, *E. servus*, in which the spot is lacking in both sexes; the male offspring had the spot weakly developed, the females were without it, and when the hybrids were inbred, some of the second-generation males had it developed to a varying degree, others lacked it altogether. These and other similar examples show conclusively that in many cases, and possibly always, each sex can transmit the secondary sexual characters proper to the other.

The problem thus arises, if each sex contains potentially the distinctive sexual characters of the other, what is it that causes the development of one set in the male and the other set in the female? In some instances, no doubt, specific "hormones" produced by the testis or ovary are the deciding factor, but the examples given in the last chapter show that this explanation is not sufficient to cover all cases. It was shown that there must be some physiological difference in the tissues themselves, but the facts of inheritance just described prove that this difference

Plate XVI

♂ *Ithysia (Nyssia) zonaria* ♀

♂ *Lycia (Biston) hirtaria* ♀

♂ *L. hirtaria* ♀ × *I. zonaria* ♂ ♀

♂
I. zonaria ♀ × *L. hirtaria* ♂

Nyssia zonaria, *Biston hirtaria*, and their hybrids from reciprocal crosses. (From Harrison and Doncaster, Journal of Genetics, III (1914), Pl. XVII.)

does not consist in the presence in one sex of hereditary " factors " for the distinctive sexual characters which are lacking in the other. No complete answer can yet be given to the problem, but some suggestions for a solution can be obtained from certain other experiments on hereditary transmission.

One of the first experiments of the kind referred to was made by T. B. Wood with Sheep, and has more recently been confirmed by Arkell and Davenport. They made crosses between two breeds of sheep, one of which is hornless in both sexes, the other horned in both sexes. Whichever way the cross was made, all the male offspring were horned, all the females hornless. The factor for horns behaves as an ordinary Mendelian unit, but is dominant in the male and recessive in the female. A male is horned if he receives the factor for horns from either parent; the female is horned only if she receives it from both parents. If H represent the factor for horns, h its absence, a male of constitution Hh will be horned, a female of the same hereditary constitution will be hornless. In individuals of this constitution (hybrids) the presence of horns is therefore a secondary sexual character, as it is in pure-bred sheep of some other races. It must be supposed that either there is something in the male which, in the presence of the factor H, causes horns to develop, or something in the female which suppresses their development except when H is received from both parents. And it is probably this " something," whatever it may be, that constitutes the physiological difference between the tissues of the male and female.

Other somewhat similar cases have been investigated in Insects. An American form of the " Clouded Yellow " Butterfly, *Colias philodice*, has two forms of the female, one with a bright yellow ground-colour like that of the male, the other nearly white; all males are yellow. Gerould found that a white female mated with a yellow male always gives both white and yellow female offspring; a yellow female may have either all yellow, or some yellow and some white female young, according to the constitution of the male with which it pairs. From the facts here shortly summarized, Gerould concludes that white is dominant over yellow in the female, but yellow over white in the male, and that white males are not produced because a white-bearing spermatozoon cannot effectively fertilize a white-bearing egg[1]. As in the sheep, a character is here dominant in one sex and recessive in the other, and Gerould suggests that the difference consists in the presence of two X-chromosomes or their equivalent in the male, and only one in the female, as was described in an earlier chapter in *Abraxas*. The presence of two X-factors would cause the yellow to appear whether it is homozygous or heterozygous (YY or Yy); in the presence of one X-factor it appears only when homozygous (YY), so that Yy males are yellow, Yy females white.

Analogous, but more complicated, cases are provided by the Papilios (Swallow-tail Butterflies) of the

---

[1] Some cases of this kind have been proved to exist, *e.g.* homozygous yellow mice cannot be produced because the zygote yellow-yellow is not viable.

Plate XVII

Papilio polytes, male and the three forms of female. (1) Male. (2) "Male" form of female. (3) Hector form. (4) Aristolochiae form.

Indian region and Africa which have three or more forms of female and only one form of male. The most completely known of these, *Papilio polytes*, investigated by Fryer in Ceylon, suggests an explanation of the same kind as that proposed by Gerould for *Colias philodice*. In *P. polytes* there are three forms of the female, one closely resembling the male ("male" form), and two others which from their resemblance to other species are commonly known by their names as the *hector* and *aristolochiae* forms (Pl. XVII). The two latter forms, which are not strikingly different from each other, have been shown by Fryer to differ in the presence or absence of a factor which is in no way connected with sex, so they may be considered together. As in *Colias philodice*, then, the female may either resemble the male or have the very distinct *hector-aristolochiae* pattern. The chief difference from *C. philodice* lies in the fact that the "male" form of the female and the *hector-aristolochiae* form may each of them have only daughters like themselves, or may have mixed families, according to their own hereditary constitution and that of their mates. Fryer suggests an explanation on the following lines: that there are two factors, one for the "male" form M, one for the *hector-aristolochiae* form H. M when homozygous (MM) is dominant over H, but is recessive to H when heterozygous (Mm). If now M is sex-limited in transmission, in such a way that it is always homozygous (MM) in males, and heterozygous (Mm) in females, (so exactly corresponding with the *grossulariata* character in the Currant Moth), all males will have the "male" form, whether they have the constitution $MMhh$, $MMHh$, or $MMHH$; females, on the other hand, will have the "male" form when their constitution is $Mmhh$, but the *hector-aristolochiae* form when it is $MmHh$ or $MmHH$[1].

These cases of dimorphism or polymorphism among the individuals of one sex may seem rather far removed from the subject of secondary sexual characters, but they have a very important bearing on the problem of their inheritance. It seems at least possible that, in these cases, the influence which determines whether the female shall resemble the male or belong to one of the other forms is a hereditary factor which is sex-limited in transmission. In the butterflies it is suggested that the male is homozygous for a factor for which the female is heterozygous, (*e.g.* that in *P. polytes*, all males are MM, all females Mm), and the presence or absence of this factor alters the constitution in such a way that the development of other inherited characters is either suppressed or allowed. These results suggest, therefore, that the difference between males and females in respect of their distinctive characters is not that one sex inherits one set of these characters while the other receives a different set, but rather that in one sex a particular set of secondary sexual characters is allowed to appear, while in the other it is inhibited, and that this development or inhibition is at least

[1] The scheme given is a modification of Fryer's, but does not differ from it in its most essential features. It should be mentioned that very similar results have been obtained from breeding experiments with *Papilio memnon*; the experiments were made by Jacobson and described by de Meijere. A still more complicated case, not yet fully investigated, is that of the African *Papilio dardanus*.

Plate XVIII

Hermaphrodites or Gynandromorphs produced by crossing *Lymantria dispar* ♀ with *L. japonica* ♂. (From Goldschmidt, by permission of the editor of *Zeitschr. f. indukt. Abstamm. und Vererbungslehre*.) (1) Male, (2) Female, of *L. dispar*. (3) Male, (4) Female, of *L. japonica*. (5–16) Hybrids, combining to a varying extent the male and female characters.

sometimes effected by a character which is sex-limited in transmission.

Further evidence, possibly pointing in the same direction, is given by the curious experiments of Brake and Goldschmidt in crossing two nearly related moths, *Lymantria dispar* aand *L. japonica* (the European and Japanese forms of the " Gypsy Moth "). The cross *japonica* ♀ × *dispar* ♂ gives normal males and females (males dark, females light); the converse cross *dispar* ♀ × *japonica* ♂ gives normal males, and females all of which have to a greater or less extent the male secondary sexual characters, not only in colour (Pl. XVIII), but also in the structure of the antennae and in the external genital organs. After inbreeding for several generations, males with flecks or patches of the female coloration were also produced. Goldschmidt suggests in explanation that the factors for the male coloration and other sexual characters are sex-limited in transmission, but that these factors in *japonica* have greater "potency" than in *dispar*, so that they appear in the females from the cross *dispar* ♀ × *japonica* ♂ even though they have been received from one parent only. He supposes that normally the male characters only appear when homozygous; but the greater "potency" of the factors for the male characters in *japonica* causes them to appear when heterozygous in the hybrid with *dispar*[1].

[1] Since this was written, a second important paper has appeared (Goldschmidt and Poppelbaum) giving the results of further experiments. When the crosses are made with a certain race of *dispar*, the gynandromorphs are not only females with some male characters, but have a genital organ consisting of both ovary and testis, and

It would be premature, to assume that secondary sexual characters, or sexual differences generally, are always produced by the action of a factor which is distinct from the sex-determining factor itself. In the last chapter evidence was given that, in some animals at least, the sexes are distinguished by recognisable physiological differences, and it seems likely that in some form or other such differences are universally present. It would probably be wrong to regard these physiological (metabolic) differences as either caused by sex or as causing sex directly and in all cases; it would rather seem that both result from the original difference in the fertilized eggs, and that in some cases (as in Insects) the physiological differences are independent of the presence of an ovary or testis, while in others (*e.g.* Mammals) the secretions of the ovary or testis largely accentuate the physiological differences. Each sex would then have the inherited potentiality of producing either male or female sexual characters, and whether one or the other set of characters appears, depends on the particular kind of metabolism of the tissues concerned[1]. This metabolism depends on various

evidence is given that larvae which were originally female may be converted during their growth not only into gynandromorphs, but actually into males. (Cf. Hertwig's "indifferent" frog larvae.) It is also shown that the gynandromorphic males which resulted from inbreeding arise only from those matings in which the female parent is of the cross *japonica* ♀ × *dispar* ♂, never from those in which the female parent is of the cross *dispar* ♀ × *japonica* ♂. For Goldschmidt's hypothesis to account for these results, the reader must be referred to the original paper.

[1] R. Hertwig [*loc. cit.* p. 138] expresses this idea by the comparison of gall-production in plants; an oak has the inherited

conditions in different cases; in Insects it appears to be a fundamental characteristic of the tissues of the two sexes; in the higher Crustacea (Crabs, etc.), while there is no evidence that it is caused by the action of the ovary or testis, it can be entirely modified in the male by the physiological disturbance induced by *Sacculina*; in Vertebrates it seems always to depend to a greater or less extent on the activity of the genital organs themselves. Lastly, in some Butterflies and perhaps in Sheep and other animals, it is possible that some differences between the sexes are controlled by a factor which has sex-limited transmission. It may be objected that the sex-limited factor suggested for such cases as *Papilio polytes* does not differ in any way from the fundamental sex-differences of other Insects; that the male form of *polytes* dominates over the *hector* form in the male, not because the factor for it is homozygous in the male and heterozygous in the female, but simply because the male physiology always causes its development, while it appears in the female only when the *hector* factor is absent. This argument is valid, unless exceptions appear, but when sex-limited inheritance of the type required is known in the same group, and when the character concerned is not properly a secondary sexual character, since it is found only in some females, it seems preferable to separate such a case from those of normal sexual

potentiality of producing a particular kind of gall when stimulated to do so by one insect, another gall when stimulated by a different insect, but the galls are only produced in response to the stimuli.

distinction, and to recognise that the explanation suggested is at least a possible one.

Finally, it follows from the facts given above that in considering the causes of the determination of sex, secondary sexual characters are not identical with sex despite their intimate relation with it. Hypotheses of sex-determination have been formulated on the basis of Mendelian inheritance of sex-determining factors; for example it has been maintained that, in some animals at least, femaleness is dominant over maleness, and that if the female-determining factor be called F, the male-determining M, females have the constitution FM, males MM. Such hypotheses will be discussed later, but meanwhile it should be noted that the fact of one sex transmitting the secondary sexual characters of the other cannot be used as a valid argument against them. The inherited factors for the secondary sexual characters may be present in each sex, and the sex-determining factor may decide which shall appear; the male might therefore transmit the female characters, even though he could not transmit the female sex-determiner. Failure to recognise this distinction has led to much unnecessary confusion and controversy, and it is therefore important that the distinction between factors for sexual characters and for sex-determination should be clearly grasped.

CHAPTER IX

HERMAPHRODITISM AND GYNANDROMORPHISM

In the preceding chapters reference has been made to several unsolved problems, and to some apparent contradictions between hypotheses which seem naturally to arise from distinct classes of facts. An attempt will be made in the next chapter to deal with the more important of these difficulties and inconsistencies, but before this is undertaken it will be convenient to discuss more fully some questions which have been mentioned incidentally, but which are of sufficient interest to deserve more detailed description.

One of the questions referred to shortly at an earlier stage, and requiring somewhat fuller treatment, is hermaphroditism—the presence of both male and female organs in the same individual. Among the lower animals, many species and some large groups are regularly hermaphrodite, and it was formerly supposed that hermaphroditism was a primitive condition, and that the separation of the sexes was a later step in evolution. This opinion was supported by the fact that most species which have separate sexes occasionally produce hermaphrodites as abnormalities, and these were regarded as reversions to a more primitive type. Further investigation shows, however, that hermaphrodite forms are commonly highly specialized; for example, in Molluscs the more primitive species are bisexual and only some of the more specialized forms, such as the land-snails and the oyster, are hermaphrodite. It is probable, therefore, that hermaphroditism must be regarded as a variation from an earlier condition in which the sexes were separate, and that the hermaphrodites which occasionally appear in normally bisexual species are produced by variations occurring sporadically in the same direction. The question then arises, whether hermaphrodites are modified females or modified males, or whether they are sometimes one and sometimes the other, or, finally, whether they are to be regarded as intermediates, combining the characters of both sexes but not derived directly from either. It is impossible at the present time to give an answer to these questions which shall be generally applicable, but there are some indications in certain species. In the Nematodes (Threadworms), for example, most species are bisexual, but some are hermaphrodite, and one at least (*Rhabdonema nigrovenosum*) has a life-history in which the hermaphrodite condition alternates with the bisexual. The hermaphrodite *Rhabdonema* lives as a parasite in the lungs of the frog; it lays eggs which are fertilized by its own spermatozoa, and these grow up into free-living males and females. Now it is found that the females and the hermaphrodites have two X-chromosomes, the males one; the hermaphrodite produces eggs all of which bear the X-chromosome,

Plate XIX

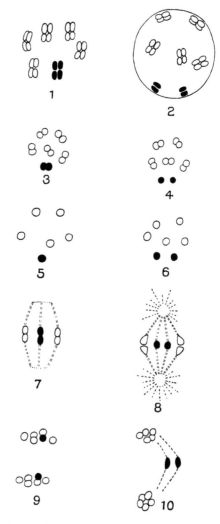

Chromosomes in the development of eggs and spermatozoa of the hermaphrodite generation of *Rhabdonema nigrovenosum*. (After Boveri, reproduced by permission of the editor, from *Quart. Journ. Micr. Science*, LIX (1914), p. 501). Oogenesis on left, spermatogenesis on right. Hetero-chromosomes black. (1, 2) Early stages of first maturation-division of egg and spermatozoon. (3, 4) Daughter groups of first maturation-division. (5, 6) Equatorial plates of second maturation-division. (7, 8) Diagrammatic side views of same stage. (9, 10) Side views of last stage of second maturation-division.

but in the production of its spermatozoa one of the two X-chromosomes which are present in the sperm-mother-cell is left out when the cell divides, so that two kinds of spermatozoa are produced, half of which have an X-chromosome and half have not (Pl. XIX). Since all the eggs before they are fertilized have an X-chromosome, and half the spermatozoa have it and half do not, it follows that after fertilization half the eggs (zygotes) have two, and become females, while the other half have one, and become males. In the next generation the spermatozoa without X degenerate, with the result that all fertilized eggs laid by the free-living females have two X-chromosomes, and all develop into hermaphrodites. In this case, therefore, it seems clear that the hermaphrodite is a modified female, but no satisfactory explanation has been given of the fact that an egg with two X-chromosomes develops into a female in one generation and into a hermaphrodite in the next.

Additional evidence that hermaphrodite Nematode worms may be regarded as modified females is given by the fact that in some species there are males and hermaphrodites, and in most cases the hermaphrodites produce spermatozoa when young and eggs when more mature. In several species the spermatozoa are stored up in the duct of the genital organ, and fertilize the eggs of the same individual when these are ready. In consequence of this, the males are unnecessary, and are frequently very rare; in several species less than one male is found per 1000 females, and not rarely the males are quite functionless and make no attempt to pair. Usually, when the store of spermatozoa is used up, all eggs produced afterwards are sterile, but in some species the eggs may develop parthenogenetically. Fräulein Krüger has described a remarkable case (*Rhabditis aberrans*) in which the egg nucleus undergoes no reduction division, and develops parthenogenetically, but a spermatozoon nevertheless enters each egg. The sperm-nucleus, however, does not conjugate with the egg-nucleus, but remains by itself and finally degenerates. It appears, therefore, that in this case the entrance of a spermatozoon may be necessary to start development, although no conjugation occurs, and the egg is in reality parthenogenetic. A number of similar cases have been observed under artificial conditions; for example, when a Mollusc spermatozoon is made to fertilize a Sea-urchin egg, the egg is caused to develop but the sperm-nucleus takes no part in the process, and the larva is entirely of the Sea-urchin type. Many such experimental cases have been described, but *Rhabditis aberrans* is hitherto the only known instance in which the same phenomena occur naturally.

The general relations of hermaphrodite to bisexual species in Nematodes described in the last paragraph would lead to the conclusion that the hermaphrodite is a modified female, but it must be admitted that in one instance at least there are facts which render this conclusion doubtful. In the parasitic species *Bradynema rigidum* the larvae are male and female; the adults are all hermaphrodite and the intermediate stages are unknown. The adults have on the whole rather the male than the female anatomical structure, and in one young individual found by Zur Strassen there was a fully developed testis with a rudimentary ovary at its posterior end. The facts as far as they are known thus suggest that in this species the female larvae degenerate and the males later develop ovaries and become the mature hermaphrodites.

In the Crustacea, reasons have been given by Geoffrey Smith for believing that hermaphrodites, when they occur, are modified males. By far the greater number of species are always bisexual, but in some forms ova may be produced in the testes, not only, as already mentioned, from the effects of the parasite *Sacculina*, but also sometimes under other circumstances. Females, on the contrary, are not known abnormally to produce spermatozoa. Other reasons for regarding hermaphrodite Crustacea as modified males have also been given by Geoffrey Smith. All the Crustacea which are normally hermaphrodite are in adult life either parasitic or permanently fixed, so that it is probable that the hermaphroditism is connected with their sessile habit. In some of the parasitic Isopod Crustacea, for example *Bopyrus*, the individuals are male when young, and become female later. When in the male condition the animal attaches itself to an older female and fertilizes its eggs; it then fixes itself to its permanent host—a prawn or other Crustacean—and there develops into a female. This, then, is a clear case of a male later becoming a definite female. In some Spider-crabs (*Inachus*) the male animal may reach a condition of sexual maturity before it is full-grown; at this stage it has large testes and well-developed secondary sexual characters. After the breeding season the testes become reduced, the claws assume the female form, and the animal undergoes a period of growth. When it has reached its full size, the testes again enlarge and become functional, and the animal reassumes the distinctive secondary sexual characters of the male, especially a conspicuous swelling of the claws. In the Sandhopper (*Orchestia*), a somewhat similar cycle is found, and in this case during the intervening growth period small ova appear in the testes, but are broken down and absorbed when the renewed activity of the testis begins.

These examples lead up to the conditions found in the Barnacles, the only large group of Crustacea which are nearly all hermaphrodite. Most of them consist of only one kind of individual which contains both male and female organs; since nearly all are gregarious, although they are sessile they usually fertilize each other. In some genera, for example *Scalpellum*, in addition to the large hermaphrodite individuals there are minute degenerate males, which live attached to the hermaphrodites just within the opening of the shell; these were named by Darwin "complemental males." Finally, in a few species, there are larger individuals which are completely female, and small males resembling the complemental males of other forms. It has usually been assumed that the hermaphrodites represent the females of the bisexual species, but the existence of the transitional forms mentioned above has led Geoffrey Smith to the belief that the hermaphrodites do not differ in their early stages from the males, and that if a free-swimming larva settles in a position in which it can grow to its full size, it passes through the male condition to that of the hermaphrodite, or even in some species becomes completely female. He regards the reduced males, on the other hand, as individuals which have fixed themselves to the females, and which, from their position, are unable to continue their development beyond the juvenile male condition[1].

Among the vertebrates, normal hermaphroditism exists only in the Hag-fish (*Myxine*), which like *Bopyrus* and some Nematodes is said to be male in its youth and to become female as it grows older. In the lower vertebrates occasional hermaphrodites are not uncommon, and, as has been mentioned in an earlier chapter, Hertwig regards his " indifferent " frog larvae as individuals which may develop into either sex, and which may give rise to hermaphrodites if neither sex completely proponderates. In the higher vertebrates true hermaphrodites are rare; so-called hermaphrodites usually have either ovaries or testes, which are often ill-developed, and, possibly in consequence of this, their sexual characters are abnormal.

[1] While this was in the press, a paper by F. Baltzer appeared (*Mitteil. Zool. Stat. Neapel*, XXII, 1914, p. 1) describing the development of sex in *Bonellia* (cf. Pl. I). He shows that as in the Barnacles, if a larva attaches itself to the proboscis of an adult female, it becomes a small parasitic male, but that all or nearly all larvae which are prevented from attaching themselves grow up into females. Gynandromorphs can be obtained by allowing larvae to attach themselves and then removing them after a short time.

If, as the study of chromosomes suggests, one sex is produced by the presence of two "sex-chromosomes" or their equivalent, the other sex by the presence of only one, it is possible that hermaphrodites arise when one of these chromosomes is for some reason enfeebled in its action. It has been suggested that the presence of this extra chromosome in the cells modifies the whole metabolism of the organism, and if for any reason this metabolism were only partially induced, it may be imagined that a hermaphrodite would result. If this is so, hermaphrodites should be modified females in all cases in which the female is characterized by two sex-chromosomes and the male by one, and should be modified males where it is the male that has two sex-chromosomes. At present our knowledge is too meagre to test whether this is so or not.

A subject somewhat akin to hermaphroditism is that of gynandromorphism, the condition which occasionally occurs in Insects, and more rarely in other animals, in which parts of the body have the characters of the male and other parts those of the female. Not infrequently, as in the Pheasant and

DESCRIPTION OF PLATE XX.

Gynandromorph Insects. (1) Apollo Butterfly, *Parnassius delius*, female on the left, male on the right side. (After Denso.) (2) Pine Moth, *Bupalus piniarius*, female on the left, male on the right side. (By permission, after Dziurzynski, *Berlin Entom. Zeitschr.* LVII. Pl. I). (3) Gynandromorph Ant (*Myrmica scabrinodis*), characters of the male on the left, those of the worker on the right. (From a photograph lent by H. Main, Esq.) (4) Egg of Currant Moth, *Abraxas grossulariata*, showing two polar spindles and (above the dotted line) two egg-nuclei conjugating with sperm-nuclei; illustrating the hypothesis of gynandromorphism suggested on p. 128. (From Doncaster, *Journal of Genetics*, IV, 1914, Pl. II.)

Plate XX

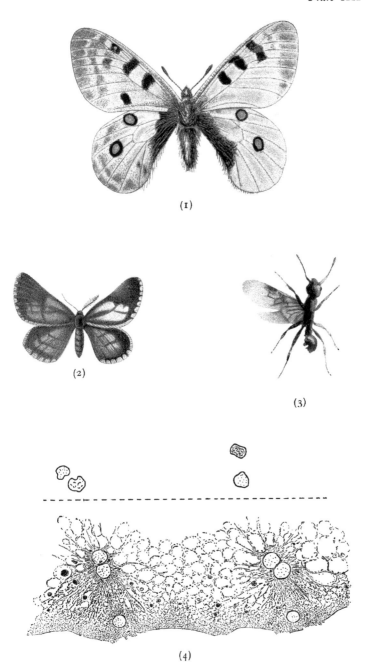

(1)

(2)

(3)

(4)

Bullfinch described in Chapter VII, the body is divided quite sharply along the middle line, the right side being of one sex, the left of the other (Pl. XX). In other cases the greater part of the body has the characters of one sex, a small portion only those of the other, or the whole animal may be made up of an irregular mosaic of parts, taken as if at random from one and the other sex. Several suggestions have been made to explain the origin of such gynandromorphs on the basis of the chromosome theory of sex-determination.

Boveri, in discussing the origin of gynandromorphic bees, suggested that the unfertilized egg-nucleus divided, as normally happens in the parthenogenetic eggs which give rise to drones, and that then the sperm nucleus conjugated with one of the two daughter nuclei so produced. There would thus be formed two nuclei, one containing the double number of chromosomes as in the normal female bee, and one containing the single number as in the male. If one of these nuclei gave rise to the nuclei on the right side of the body, the other to those on the left, a bilateral gynandromorph would be produced. Morgan has suggested a different origin; he points out that two or more spermatozoa often enter an insect egg, and although usually only one of these takes part in fertilization and the rest degenerate, he regards it as possible that sometimes a spermatozoon may give rise to a nucleus which would take part in the production of the embryo without conjugation. Such a sperm nucleus, which has not conjugated, might be supposed to determine male characters, while the zygote-nucleus formed by union of the egg nucleus with a second sperm nucleus would give rise to parts having the female character[1].

Both these hypotheses have two defects. Firstly, they do not easily account for the fact that gynandromorphs are sometimes especially abundant in certain families (cf. Pl. XXI). Hives of bees have been recorded in which gynandromorphs were frequent, while in other hives they are non-existent. This suggests that the condition is due to some peculiarity in the egg, since all the bees of one hive are in general the offspring of one queen. It is possible, of course, that in the eggs of certain queens the nucleus may tend to divide prematurely, but there is no direct evidence for this. Secondly, neither hypothesis will easily account for the existence of gynandromorphs in animals in which parthenogenetic division of the egg-nucleus is unknown, as in Butterflies. The writer has recently observed a condition in the Currant Moth (*Abraxas*) which suggests a third possible origin. He found that in some eggs there were two nuclei, both of which undergo maturation simultaneously, and both conjugate with sperm-nuclei, so producing two zygote-nuclei in one egg (Pl. XX, fig. 4). If in the maturation

[1] Lang has suggested a combination of these two possibilities—that the egg-nucleus divides prematurely and that one half is fertilized by a male-determining spermatozoon, the other by a female-determining. He also suggests that a bilateral gynandromorph might arise by an abnormal division of the fertilized egg-nucleus, of such a kind that a sex-chromosome, instead of dividing normally, is caused to go undivided into one daughter-nucleus. A similar hypothesis has quite recently been adopted by Morgan (*Proc. Soc. Exp. Biol. and Med.* XI. 1914, p. 171).

Plate XXI

Part of a family of *Amphidasys betularia* ("Peppered Moth") which included eight gynandromorphs in addition to 20 males and 40 females. Three gynandromorphs had the male antenna on the left, four on the right (six upper figures), and one had intermediate, slightly feathered antennae (middle figure). The two lower figures represent normal male and female. From a photograph lent by H. Main, Esq. Described in *Entomologist*, XXXIV. (1901), p. 203.

Plate XXII

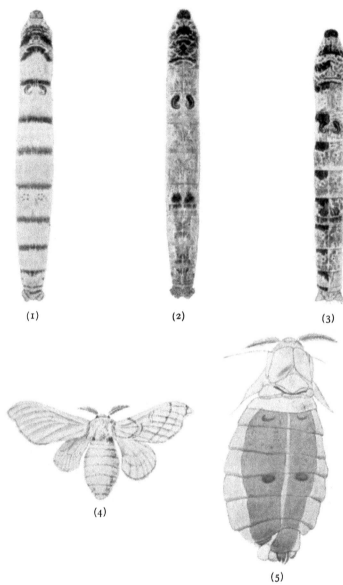

Gynandromorph Silkworm, *Bombyx mori*, produced by crossing a female of a race with striped larva by a male of a race with unstriped larva. (From Toyama.) (1) Normal striped larva. (2) Normal unstriped larva. (3) Gynandromorph hybrid larva, striped on left side, unstriped on right. (4) Moth reared from larva represented in fig. 3, female on left, male on right side. (5) Enlarged view of the body of the same moth.

divisions one of these nuclei expelled the sex-chromosome in the polar body while the other retained it, one of the nuclei of the fertilized egg would be male-determining and the other female-determining, and a gynandromorph might be expected to result. Since two of the three observed cases of binucleate eggs were laid by one female, of which only 45 eggs were available for examination, while only one other case was found among hundreds of eggs of other females, it may reasonably be inferred that a tendency to the production of the binucleate eggs is present in certain females, and so the tendency of gynandromorphs to appear in certain families would be accounted for.

That this is perhaps not the only possible cause of origin of gynandromorphs is indicated by two specimens described by Toyama in the Silkworm (Pl. XXII). He crossed a female of a race with striped larvae with a male of an unstriped race. Two of the resulting larvae were striped on the left side of the body, unstriped on the right. When these were reared to moths, it was found that both in the secondary sexual characters and the external genital organs the left side was female, the right side male. The female side of the body thus inherited the larval characters of the mother, the male side those of the father. This raises some very puzzling problems. The striped condition is dominant over the unstriped, so that if the factor for striping is present it should show itself; it must therefore be concluded that the right side of the body did not receive this factor from the mother. On the other hand, it was shown in

Chapter V that in Moths the female apparently has one sex-chromosome, the male two, and this is supported by the fact that in the rare cases when Moths' eggs develop parthenogenetically, they usually, if not always, produce females. It might be suggested from this that the left side was derived from an unfertilized egg-nucleus, the right side from a fertilized nucleus, but if this was so, it is difficult to understand why no striping appeared on the right side, since the egg-nucleus must have borne the factor for striping. Two explanations seem possible. The first is that the female parent was heterozygous for striping, and the gynandromorph arose from a binucleate egg of which one nucleus bore the striping factor and the other did not; this is unlikely, for Toyama's account, while not perfectly definite on the point, implies that the female parent belonged to the pure striped breed. The second suggestion is that the gynandromorph may have arisen by an abnormal division of the fertilized egg-nucleus, of such a kind that one sex-chromosome divided normally, while the other went undivided to one pole, and concurrently the chromosome which is concerned with the transmission of the striping factor went undivided to the other pole. If Y represents the sex-chromosome, and S the factor of striping, the egg-nucleus before fertilization would be YS, the sperm nucleus $Y-$, and the fertilized egg $YYS-$. If now at the first division of the fertilized nucleus one Y divides, while the other Y and S go to opposite poles, thus

$$Y \leftarrow\!\rightarrow Y$$
$$S \leftarrow\!\rightarrow Y$$

one of the daughter nuclei will contain YY and no S, and have the factors characteristic of an unstriped male, as in the right side of the gynandromorph; the other daughter nucleus will have one Y and S, and have the characters of a striped female, as in the left side. In this connexion it may be mentioned that a few instances have been recorded of moths in which the two sides of the body have the characters of different varieties, although both sides are of the same sex. An example of this condition is described by Standfuss in *Aglia tau*, and another figured by Stichel in *Lymantria monacha*. In the case of *A. tau* one side had the typical coloration, the other that of the variety *lugens*, which is known to behave as a Mendelian dominant with respect to the typical form. This instance may thus be compared with Toyama's Silkworm, from which it differs in the fact that both sides of the body were female, instead of being also gynandromorphic. Comparable cases have been figured by Morgan in the fly *Drosophila*.

One other subject, not very closely connected with those just discussed, may conveniently be mentioned here. This is the tendency sometimes found in certain families, to produce offspring all or nearly all of one sex. A curious case of this has been described by Federley, who obtained in two successive generations broods consisting wholly of females in the "Chocolate-tip" Moth, *Pygaera pigra*. In this instance he found that the lack of males was due to an inherited disease, which caused the blood of the male larvae to become abnormal, and killed off all the males in the larval or egg stages; females were unaffected, although they transmitted the disease to their male offspring. This case provides another interesting example of the physiological difference between the two sexes of one species. In other instances of an inherited tendency to produce families consisting entirely or very largely of females no such explanation is possible; the writer has bred seven generations of a strain of *Abraxas* in which some broods in each generation are all females, and an analogous case has been described by Poulton in the African Butterfly *Acraea encedon*. In these cases the tendency appears to be transmitted in the female line, and there is no evidence that the lack of males in certain broods is due to their death in the early stages. It is possible, but not proved, that in these families all the eggs have the tendency to extrude the male-determining sex-chromosome, just as in Aphids the male-producing eggs all extrude the female-determining X-chromosome.

Other curious cases of the same sort have been described in certain hybrids between different species of Moths; for example, in crosses between the Moths *Biston hirtaria* and *Nyssia zonaria* already referred to on p. 110, if *hirtaria* is used as the female parent, both males and females appear in the offspring, but when a *zonaria* female is mated to a *hirtaria* male, all the offspring are males (Pl. XVI). It is possible that this is a more extreme case of the same kind as that of *Lymantria dispar* and *L. japonica*, described in Chapter VIII; when a *japonica* male is mated with a *dispar* female both male and female offspring have the male secondary sexual characters, and this may be regarded as a partial conversion of the female into a male, perhaps, as Goldschmidt suggests, by the greater "potency" of the male characters in *japonica*. In the cross *zonaria* ♀ × *hirtaria* ♂ the same process may perhaps be carried further, and cause eggs, which with sperm of their own species would have given females, to produce not only male secondary sexual characters, but actual males. Put quite crudely, a *zonaria* egg may be supposed to require a certain amount of male-determining substance to make it become a male, and when fertilized by a *zonaria* spermatozoon this only happens if the egg already contains a male-determining factor; the *hirtaria* spermatozoon, on the other hand, may be supposed to contain so much "male-determining substance" that it can cause any *zonaria* egg to become a male, whether the egg contains a male-determining factor or not. This hypothesis, though speculative, is suported by the fact that most of the *hirtaria* chromosomes are about four times as large as those of *zonaria*, so that, if one of them may be regarded as consisting of "sex-chromatin," any *zonaria* egg fertilised by *hirtaria*, whether it was originally male-determining or not, would contain enough sex-chromatin to cause it to become a male.

In this connexion, and also as showing that in some cases at least the production of families which are all of one sex is not altogether unrelated to the subject of gynandromorphism, reference may be made to an interesting discovery of J. W. H. Harrison. He made reciprocal crosses between *Biston hirtaria* and *Poecilopsis pomonaria*, a moth somewhat related to *N. zonaria*. Whichever way the cross was made, both males and females were produced, the females having small wings as in the cross *zonaria* ♂ × *hirtaria* ♀. When the male hybrid from the cross *pomonaria* ♀ × *hirtaria* ♂ was mated back with a *hirtaria* female, again males and females were produced, but when the female hybrid was mated with a *pomonaria* male, all the offspring were gynandromorphs.

These results may be represented thus:

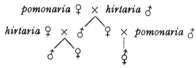

The gynandromorphs might almost be called hermaphrodites; they are not divided along the middle line, but show a mosaic of male and female features, and their external reproductive organs show a remarkable combination of the structures characteristic of the two sexes. This case, when compared with the result of the mating *zonaria* ♀ × *hirtaria* ♂, in which only males are produced, tends to confirm the conclusion that, at least when distinct species are crossed, the sex of the offspring does not depend simply on the presence or absence of a sex-determining factor, but on the relative intensities of the factors introduced from the two parents. Since fertile eggs were obtained from only one mating, and these were only a small proportion of the eggs laid, it would be fruitless to attempt a detailed analysis of the case, for it is not known whether if more eggs could have been reared to maturity males or females might not have appeared in addition to the gynandromorphs. But if it may be assumed that in all moths sex is determined as it appears to be in *Abraxas*, that is, that males arise from fertilized eggs which have two " sex-chromosomes," females from those which receive the sex-chromosome from the male parent only, then these gynandromorphs may perhaps be regarded as arising from eggs which have two sex-chromosomes, but in which the *pomonaria* chromosome is not sufficiently intense in its action to transform the moth completely into a male. It must be admitted, however, that until the case has been investigated more completely, any such explanation must be purely speculative.

CHAPTER X

GENERAL CONCLUSIONS ON THE CAUSES WHICH DETERMINE SEX

BEFORE attempting any discussion of the relations to each other of the various lines of evidence sketched in the preceding chapters, it will be well to summarize the more important conclusions to which they have led already. Beginning with the question of the stage at which sex is determined, it was shown that in some cases it appears to be determined already in the unfertilized egg, in other cases to depend on the spermatozoon and to be fixed at fertilization, and in other cases again to be capable of modification during the embryonic development or even at a later stage.

Evidence for determination by the egg apart from fertilization was drawn (1) from the facts of parthenogenesis, (2) from sex-limited transmission by the female, and (3) from the cases in which two kinds of fertilizable eggs are produced, which differ from each other in their chromosomes. In all these cases it is certain that, normally at least, male-determining and female-determining eggs are produced, and that if the kind of egg is known, the sex can be predicted without reference to the spermatozoon.

Similarly, evidence for sex-determination by the spermatozoon is provided (1) by the cases in which unfertilized eggs yield males, fertilized eggs females, as in the bee, (2) by sex-limited transmission by the male, and (3) by the existence of two kinds of spermatozoa differing in respect of their chromosomes. In each of these three groups the evidence for sex-determination by the spermatozoon is exactly comparable with the similar evidence for determination by the egg in the previous class.

If the facts here shortly referred to were taken alone, the problem of sex-determination would be fairly clear. Sex might be regarded as depending on the presence of a greater or less amount of some " sex-determining substance " present in the chromosomes, or more correctly, on the physiological condition arising from the interaction of this substance with the rest of the substances of the cells. The presence of an additional " dose " of this substance in a cell otherwise similar would alter its metabolism (*i.e.* general physiological condition), and since all cells of the body would contain the extra dose, the whole physiology of the body would be affected, and the sex of the animal would be irrevocably determined. In some groups of animals the required physiological condition would be produced by the presence of two sex-chromosomes in the female cells and only one in those of the male; in other animals, owing doubtless to some difference either in the chromatin or in the cell-substance, the converse arrangement would be found—that of two sex-chromosomes in the male and one in the female.

To the first group belong most Insects, the Myriapods (Centipedes, etc.), Spiders, Nematode worms and at least several Mammals, in all of which the female always has two sex-chromosomes and the male often has only one. To the second group belong the Lepidoptera (Butterflies and Moths) and probably Birds.

So far the problem is relatively simple; although nothing is known of the manner of action of the sex-determining factor supposed to reside in the sex-chromosomes, it can at least be said that in the cases mentioned it is inherited like any other Mendelian character (as was first suggested by Bateson and by Castle), and that individuals which receive it from both parents would be of one sex, those to which it is transmitted by one parent only, of the other sex.

To this scheme, so attractive in its comparative simplicity and its close accord with the facts on which it is based, there are opposed a series of observations, usually derived from special cases and differing widely in kind among themselves, any one of which might perhaps be regarded as due to error or to chance, but which, when taken together, make a rather formidable obstacle to the acceptance of the hypothesis. They may be grouped under two heads, including (1) evidence that the egg may influence the sex in cases in which observations on chromosomes indicate that the sex should be determined by the spermatozoon; and (2) evidence that the sex may be modified after fertilization by influences acting on the embryo or even later in life.

General Conclusions

Of the first group of difficulties, several instances were given in the chapter on the sex-ratio. For example, there is general agreement among those who have investigated the causes of the fluctuations of this ratio in Man that these fluctuations are not caused by influences acting on the male parents only. Although there is wide disagreement among different writers about the nature of the disturbing causes, they generally conclude that these causes act at least as much on the mother as on the father. And yet the cytological evidence points directly to sex-determination in Man by the spermatozoon, for the spermatozoa are known to be of two kinds, both from the facts of sex-limited inheritance and from the observations of von Winiwarter on the spermatogenesis.

More direct evidence in the same direction is furnished by Hertwig's experiments with frogs. He maintains that over-ripeness of the eggs causes a preponderance of male offspring which cannot be explained by selective mortality, and which is independent of the male parent, and yet, in the same species, he found that some males produce a much higher percentage of female offspring than others, when mated with any female. He concludes therefore that both egg and spermatozoon influence the sex, a conclusion irreconcilable with the hypothesis that two kinds of germ-cells are produced by one or other sex, but not by both. In the frog there is no evidence to show whether either eggs or spermatozoa are of two kinds, but in either case the chromosome hypothesis in its simple form cannot be of general application if Hertwig's conclusions are correct.

To escape from difficulties of this sort, and also from the seeming improbability that in one group of insects and of vertebrates sex should be determined exclusively by the egg, and in other groups of the same classes of animals exclusively by the spermatozoon, it has been supposed that in reality both eggs and spermatozoa are of two kinds. It is suggested, for example, that in those species in which there are two sex-chromosomes in the female and only one in the male, the two of the female are unlike, one bearing a male-determining factor and one a female-determining, while the single one of the male is male determining. If, then, M stands for male determiner, F for female determiner, and f for absence of sex determiner, a female would have the constitution MF, femaleness being dominant over maleness, and would produce two kinds of eggs, M and F; a male would have the constitution Mf, and produce two kinds of spermatozoa, M and f. In the groups in which the female has one chromosome less than the male, the male would be MF, the female mF, M being dominant over F. Such schemes have the grave disadvantage that they involve the hypothesis of selective fertilization, that is, for example, that M-bearing spermatozoa can only fertilize F-bearing eggs, and f-bearing spermatozoa fertilize M-bearing eggs, for otherwise the combinations MM and Ff would be produced, and these are never found in practice. No direct evidence in favour of selective fertilization exists, and until it is found, the scheme must be regarded as hypothetical. It has the further objection that it is inconsistent with the hypothesis that the factors for characters which have sex-limited transmission are borne by the same chromosomes that carry the sex-determiner, for if the red-eye factor in *Drosophila*, for example, were transmitted by the male only to his female offspring because they are borne by the chromosome which also bears the factor M, then for the same reason they should be transmitted by the female only to her sons, and this does not occur. This objection has, perhaps, less weight than the last, for the evidence that factors for sex-limited characters are borne in the sex-chromosome is by no means conclusive. Apart from these specific objections, it must be admitted that any scheme of separate male and female determiners is so much more complicated and artificial than the simple one of presence or absence of one factor, that it is perhaps preferable, in the present state of our knowledge, to seek some other way out of the difficulty rather than to adopt any hypothesis of the type of that just described.

It seems certain, however, unless the alleged modifications of the sex-ratio by environment are rejected as due to error or to chance, that the simple hypothesis of an unchangeable hereditary entity, the presence of which always causes one sex and its absence the other, must be given up. If once it is admitted that sex-determination does not depend on an unmodifiable unit, but rather on the reciprocal action between an inherited factor and its surroundings, there are a number of indications which point towards a way of escape from the dilemma. In the Nematode *Rhabdonema*, for example, the presence of a particular chromosome in one generation seems to cause the individuals which contain it to be female, while in the next generation, the same chromosome grouping is characteristic of hermaphrodites. In the cross between the Moths *Nyssia zonaria* ♀ × *Biston hirtaria* ♂, although on the analogy of other Moths it must be supposed that there are both male and female-determining eggs, all the offspring are males. In the Crab, the physiological condition induced by the parasite *Sacculina* goes far towards converting a male into a female, and if Steinach's observations are to be accepted, the same sort of change may follow in Mammals from transplanting ovaries or testes into the other sex. All these facts suggest that the presence of a particular chromosome provides only one side of a reciprocal reaction; the sex depends on the complete result. If the balance is hanging evenly, the addition of the extra chromosome will cause it to incline decisively to one side, but if the balance is weighted already in some other way, this effect may not follow.

The chromosome hypothesis depends almost entirely on two classes of observations; primarily, on the many cases already known in which a chromosome is present in one sex and completely absent in the other, and secondarily on the facts of sex-limited inheritance, the best known examples of which occur in animals which have been shown to have one more chromosome in one sex than in the other. But in many animals, probably the great majority, there is no visible difference between the chromosomes of the male and female; the chromosomes of the two sexes are to all appearance alike, and it is only inferred, from the analogy of the cases where one sex has an unequal and the other an equal pair, that in one sex there is a pair of physiologically dissimilar, in the other a pair of similar chromosomes. This inference is strongly supported by the fact that in closely related species, or even in strains of the same species (*e.g. Abraxas grossulariata*), one may have an unpaired chromosome, and the other a pair of chromosomes which appear identical, in the same sex. For all that is known to the contrary, however, there may be a continuous series in different species from cases in which the chromosomes of the male and female are nearly identical, not only in appearance but in function, through cases in which they are more and more dissimilar, to the extreme case of the presence of a chromosome in one sex which is completely absent in the other. The last condition is especially characteristic of the Insects, a group in which efforts to change the sex-characters by transplantation of ovaries and testes, or to influence the sex-ratio by external conditions, have been peculiarly unsuccessful. In the Frog, on the other hand, there is no evidence of any distinction between the chromosomes of the two sexes, and it is not known whether the male or the female is the sex which produces two kinds of germ-cells. It is possible, therefore, that it belongs to the class in which the chromosomes of the two sexes are so nearly alike that a relatively small change may so alter the reaction between chromosomes and cell-substance that an egg which would have given rise to one sex is caused to produce the other. This was the conclusion arrived at by Hertwig from his observations on "indifferent" larvae in the Frog; he suggests that eggs which with normal spermatozoa would have given males and females may become indifferent when fertilized by an "indifferent producing" male, and that "indifferent" eggs, fertilized by the same male, may be converted into females. The conclusion from such facts seems to be that in the frog the balance between male and female is very evenly poised, and may be inclined to the male or female side by a variety of circumstances. In the Insects, on the other hand, the difference between the presence and absence of the sex-chromosome is so great, that no ordinary influences have any effect in comparison with it; if the chromosome is present, the balance inclines decisively to one side, if it is absent, to the other.

The general conclusion would thus be that sex is dependent on a physiological condition of the organism, a condition depending on the interaction of certain chromosomes with the protoplasm of the cells, and therefore determined, in the absence of other disturbing factors, by the presence or absence of these particular chromosomes. If the difference between the chromosomes of the male and female is considerable, it will outweigh any other influences which might tend to affect the general result;

every cell of the body will have either the male or the female condition, and no external agency will be able to affect either the sex or the secondary sexual characters. This condition is especially characteristic of the Insects. When the difference between the chromosomes of the two sexes is less, but still sufficient to outweigh the effects of most environmental changes, the difference will usually be sufficient to turn the scale decisively to one sex or the other, but the secondary sexual characters will be less dependent on inherent differences in the tissues of the animals, and more on the influence exerted by the secretions of the sexual organs. This is the condition found in Birds and perhaps less markedly in Mammals. Finally, when the chromosome-differences between the sexes are still smaller, they will only be able by themselves to determine the sex when no other causes influence the chromosome-protoplasm relation; if this relation is affected by other agencies, it becomes possible for an egg which would otherwise have been female to develop into a male. This nearly evenly-balanced condition is best known in the Amphibia, but there are indications that it is approached by some mammals.

A hypothesis of this kind avoids the more serious difficulties incurred by those which assume sex to be determined in all cases by an unchangeable factor borne by a chromosome. It is evident in any case that sex cannot depend on a chromosome alone, for the chromosome must act by its relation with the cell-protoplasm, and it is on this relation that the sex-determination depends. A relation must have two sides; it inevitably suggests to the mind the idea of a balance, and if sex-determination is regarded on the analogy of the inclination to one side or the other of a balance, it is clear that it will depend not only on whether a particular weight is placed on one side, but also on the presence or absence of other weights on one or both sides. The hypothesis suggested supposes that the sex-chromosomes correspond with a pair of weights which may be nearly the same or very different, placed on opposite sides of the balance; when the difference between them is large, the heavier will always cause the scale to fall on its side, but when they are nearly equal, other disturbing causes may sometimes cause the scale to sink on the side on which the lighter chromosome-weight has been placed. Put in different words, every germ-cell would bear a sex-determining factor, but when this factor has relatively small intensity of action, its effect may be counterbalanced by other causes which alter the physiological relation on which sex-determination depends.

CHAPTER XI

THE DETERMINATION OF SEX IN MAN

THE main purpose of this little book is to discuss the nature of the causes which lead to the production of one sex or the other, and the various special cases which have been described have been used rather as illustrations of general principles than as being of particular interest in themselves. It is the aim of science to discover principles of this general character, but they are apt to seem somewhat lacking in interest if it cannot be seen how they are applicable to those special cases which affect us most nearly. It may therefore be justifiable to consider to what extent the conclusions arrived at in the last chapter answer the questions asked at the beginning of the first. It must be admitted that at the present time they can give very little help in providing answers with regard to individual cases, and that hitherto but little progress has been made in the direction of predicting the sex of any child, and, if possible, even less in artificially influencing the determination of its sex. When the general principles arrived at are borne in mind, it must be confessed that the prospects of our ever attaining this power of control or even of prediction are not very hopeful, but the possibility of it cannot yet be regarded as entirely excluded.

The general conclusions arrived at were that sex is determined by a physiological condition of the embryonic cells, that this condition is induced, at least in the absence of disturbing causes, by the presence or absence of a particular sex-chromosome, but that there is evidence, which for the present at least cannot be neglected, that certain extraneous conditions acting on the egg or early embryo, may perhaps be able to counteract the effect of the sex chromosome. Quite generally, then, there are two conceivable methods by which the sex might be artificially influenced in any particular case; firstly, if means could be found of ensuring that any particular fertilized ovum received the required chromosomes, and secondly, by the discovery of methods which always affect the ovum or embryo in such a way as to produce the desired sex. Many suggestions for applying both methods have been made, some of which have attained considerable notoriety, but hitherto none of them has stood the test of practical experience.

In the case of the higher animals, especially of the mammals, in which the embryo develops in the maternal uterus until long after the sex is irrevocably decided, it is obviously difficult to apply methods which might influence the sex after fertilization, even if it were certainly known that such methods were ever really effective. As has been seen in the preceding chapters, apart from a few experiments

The Determination of Sex in Man

like those of Hertwig on rearing tadpoles at different temperatures, there are very few cases in which there is even a suggestion that the sex of the fertilized egg can be modified by environment, and the belief that this is possible has been entirely abandoned by many of the leading investigators of the subject. It is probable, therefore, that if it will ever be possible to predict or to determine artificially the sex of a particular child, the means will have to be sought in some method of influencing the output of germ-cells in such a way that one kind is produced rather than the other. It is in this way that Heape and others interpret the results of their investigations; they find that certain conditions affect the sex-ratio, and they explain the result by assuming that under some circumstances male-determining ova are produced in excess, under other circumstances female-determining.

Here, however, a serious difficulty arises. It was mentioned in connexion with the relation between chromosomes and sex that Man has been described as one of the species in which the male has one chromosome less than the female; all the ova contain an X-chromosome, while half the spermatozoa have it and half have not. It is true that different observers are not in agreement on the matter. Guyer found a double X-chromosome in the male, with a total chromosome number of 22, giving two classes of spermatozoa with 12 and 10 respectively; more recently, von Winiwarter found 47 in the male, giving spermatozoa with 23 and 24, and probably 48 in the female, so that all eggs would have 24[1]. From these discordant results, neither of which, it may be mentioned, is in agreement with the observations of Gutherz, it would perhaps be rash to conclude that man belonged to the class with two sorts of spermatozoa and with all eggs alike in respect of their chromosomes, but there are other facts which make such a conclusion almost unavoidable. The most important of these is the existence in Man of sex-limited transmission by the male, as seen in the inheritance of colour-blindness, night-blindness and haemophilia. If a man transmits certain characters always or nearly always to his daughters, the conclusion can hardly be avoided that he produces two kinds of germ-cells, female producing which bear the factor for these characters, and male-producing which do not. Indirect, but confirmatory evidence in the same direction is provided by the discovery of an unpaired X-chromosome in the male of other Mammals, for example, the the Guinea-pig (by Miss Stevens), the Opossum (by Jordan), and the Pig (by Wodsedalék).

If, then, sex is determined simply by the presence or absence of an additional X-chromosome, it must depend in Man entirely on the spermatozoon, and the belief in the existence of male and female-determining ova must be regarded as illusory. The only other possibilities appear to be, either that there are not only two kinds of spermatozoa, but

---

[1] Guyer's observations were made on the spermatogenesis of a negro, von Winiwarter's on that of a European. It has been suggested that the discrepancy is due to differences between these two races of mankind.

The Determination of Sex in Man

also two kinds of ova, or that the ovum may be so modified under certain circumstances as to counteract the effect of the X-chromosome in the spermatozoon. Both these hypotheses involve serious difficulties, but the subject is of such general importance, apart from its purely human interest, that some further discussion of the matter may be desirable.

The more important facts which bear on the problem have already been given in the chapter on the sex-ratio, and only a short summary of them is needed here. From Hertwig's work on Frogs it was concluded that both the male and female parents can influence the sex-ratio, and it was pointed out that this conclusion, if substantiated, was inconsistent with the belief that sex is invariably determined by the existence of two kinds of germ-cells in one sex or in the other, but not in both. Hertwig showed further that prolonged postponement of fertilization caused an enormous preponderance of males, and analogous results were obtained by H. D. King; Hertwig interprets his result by assuming that the over-ripe egg tends to undergo the maturation process in such a way as to cause the egg to be of the male-producing type. If the frog belongs to the class in which there are two kinds of mature eggs and only one kind of spermatozoon, such an explanation is possible, but unfortunately no definite data are available for deciding whether this is so or not. There is one observation of some importance in this connexion; Loeb found that when frogs' eggs were made to develop without fertilization by artificial means, the only larvae (two) which grew old enough to allow of an examination of the sexual organs showed ova in the genital gland, but he points out that these were not certainly females, for they had the characteristics of Hertwig's "indifferent" larvae which may turn into males. As far as it goes, however, the observation would suggest that when the frog's egg develops with no sex-chromosome derived from the male it has at least a female tendency, and that therefore the frog possibly resembles the moths in having one sex-chromosome in the female, two in the male. If this is so, Hertwig's explanation of his result may be sound, for if over-ripe eggs tended to retain the sex-chromosome at maturation, when fertilized they would develop into males.

If, however, we are justified in assuming that all mammals belong to the other class—that in which the spermatozoa are of two kinds—Hertwig's explanation will not apply to Pearl's observation that in Cattle, as in the Frog, over-ripeness of the ova tends to produce an excess of males. Nor can it apply to the fluctuations of the sex-ratio with season or other conditions in Man, and in Dogs and other animals. It is hardly possible to ascribe these fluctuations to changes in the proportion of male- and female-determining spermatozoa, since, unless one kind of spermatozoon degenerates, the two sorts must be produced in equal numbers, in consequence of the origin of two of each kind by the division of one cell. It is of course conceivable

XI] *The Determination of Sex in Man* 153

that seasonal and other changes might, perhaps, alter the ratio of effective male- and female-determining spermatozoa, but this would not account for those changes in the sex-ratio which appear to depend on influences acting on the mother only. Some of these, such as the age of the mother, have been referred to already, but one other hypothesis, the value of which it is difficult to estimate, remains to be mentioned. This hypothesis, which has been worked out most fully by Rumley Dawson, assumes that male- and female-determining ova are discharged alternately from the ovaries. In woman one ovum is usually discharged each month, and it is maintained that in one month the ovum is male-determining, in the next, female-determining. It is obvious that exceptions must occur, for boy and girl twins are quite common, but if the cases which support the hypothesis are taken by themselves, and the exceptions explained away, it is possible to make out a strong case in favour of the belief.

By those who hold it, it is usually further assumed that the right ovary produces male-determining ova, the left ovary female-determining, and that the two ovaries discharge an ovum alternately, but an impartial examination of the evidence for this belief shows that it rests on very slender foundations. It is certainly not true of such mammals as Rats, which produce several young at a birth, for it has been shown repeatedly that after the complete removal of one ovary, such animals may produce young of both sexes. In human beings so many instances have been recorded in which a woman has produced children of a particular sex after the corresponding ovary has been removed that it is hardly possible to believe that the removal in all these cases was incomplete. The belief in a monthly alternation is more difficult to test, for it involves the knowledge of the regularity of the discharge of ova from the ovaries. If ova are discharged very regularly at intervals of 28 days, there are 13 such periods in the year, and if a boy is born in a particular month this year, a child born in the same month next year should be a girl, one in the same month two years later, a boy. In some families, especially when the children are born at short intervals, there is remarkable concordance with expectation, but in others there are as many exceptions as correct results. But it should be noted that if the period is on the average one day early or late each month, it will be about 26 days wrong after two years and so will throw the calculation out by a month; it thus follows that before the hypothesis is tested in any given case, it must be known whether the discharge of ova is regular, and whether it has an accurate 28-day period. On the whole, it must be concluded that the belief is insufficiently supported by the evidence, and that it possibly rests on chance successes in prediction in some individual cases.

The same conclusion must probably be drawn with regard to the several attempts which have been made to influence the sex by specific nutrition of the mother. One of these which attracted great attention at the time of its publication was that of Schenk, in 1898 and 1901. He based his method

XI] *The Determination of Sex in Man* 155

on the observation that a number of women whose children were all girls all excreted sugar, as happens in people affected with diabetes. From this he suspected that the physiological condition which leads to the excretion of sugar was inimical to the development of male-determining ova, and that males could be produced by its prevention. He therefore recommended that those who desire a male child should undergo treatment similar to that prescribed for diabetes for two or three months before conception, and supposed that thus a boy would be produced. Although his treatment had considerable vogue at the time, it was found not to be effective, and it seems clear that it rested simply on the chance that the cases examined had a tendency to diabetes, and that his theory has in reality no genuine foundation.

The general conclusion with regard to man must therefore be that if sex is determined solely by the spermatozoon there is no hope either of influencing or of predicting it in special cases. On the other hand, there is considerable evidence, admittedly not of a very conclusive character, that the ovum has some share in the effect, and if this is so, before any practical results are reached it will be necessary to discover which of two conceivable causes of sex-determination is the true one. It is possible that there are two kinds of ova as well as two kinds of spermatozoa, and that there is selective fertilization of such a kind that one kind of spermatozoon only fertilizes one kind of ovum, the second kind of spermatozoon the second kind of ovum. If this should prove to be the case, it is possible that means might be found of influencing or predicting the kind of ovum which is discharged under any particular set of conditions. Secondly, it is possible that the ova are potentially all alike, but that their physiological condition may under some circumstances be so altered that the sex is determined independently of the spermatozoon. This latter hypothesis, as explained in the last chapter, avoids the serious difficulty that the selective fertilization involved in the former is not known to exist, but it incurs the almost equally grave difficulty which arises from the facts of sex-limited transmission. If a male nearly always transmits a character only to his daughter, it would seem that the ovum cannot have more than a negligible share in the determination of the sex. In the case of man, however, there is one point briefly referred to earlier which may have some bearing on this part of the subject. Most pedigrees of sex-limited affections in man show a considerable excess of males in the families which include affected members. This excess of males is seen especially in pedigrees of haemophilia and night-blindness. To such families the affection is transmitted by the mother, and hence, if the excess of males is genuine, it must depend on the mother rather than on the father. It has been maintained that the apparent excess of males is due not to a real excess in such families, but to the fact that a family is only known to be affected when it includes at least one affected male, and that therefore families which have a large proportion of daughters are omitted from the statistics and so an excess of males appears. In

XI] *The Determination of Sex in Man* 157

the apparently analogous case of the tortoiseshell cat, however, the statistics collected by the writer show an even greater excess of males from the corresponding mating (67 males, 35 females, see p. 48), although in this case the families were selected not by the appearance of a character among the offspring, but by the character of the mother, so that in this case there seems no reason to suspect any statistical fallacy, apart from the smallness of the numbers.

With our present knowledge it does not seem possible to construct any detailed scheme of a probable nature to explain the facts, and, as has been mentioned above, it is not even certain that the excess of males in families affected by these sex-limited diseases is genuine. If it is, it is hardly possible to avoid the conclusion that the sex of the offspring may be influenced, at least under certain circumstances, by the mother, since in the cases mentioned the father is normal and the abnormality is transmitted by the mother only. If a mother who transmits such a disease as haemophilia tends to produce a preponderance of sons, although at present we have no knowledge whatever of the manner in which the change in the sex-ratio is brought about, it seems at least possible that other physiological conditions of the mother may also influence the sex-ratio. From this it would follow that the search for means of influencing the sex of the offspring through the mother is not of necessity doomed to failure. No results of a really positive kind have been obtained hitherto, and some of the facts point so clearly to sex-determination by the male germ-cell alone in Man and other Mammals that many investigators have concluded that the quest is hopeless; but until an adequate explanation has been given of the phenomena mentioned in the preceding paragraphs, it seems more reasonable to retain an open mind, and to regard the control of sex in Man as an achievement not entirely impossible of realization.

GLOSSARY OF TECHNICAL TERMS

The pages are those on which the use of the term is explained in the text.

Accessory Chromosome, *see* Heterochromosome.

Cell. A cell has been defined as "a unit mass of living matter." It consists of a portion of protoplasm (living substance) containing a nucleus; all living things are built up of cells.

Chromatin. The substance which is especially characteristic of the nucleus of the cell; it stains more easily and intensely than the other constituents of the nucleus. (pp. 9, 50.)

Chromosome. A portion of chromatin which in nuclear division behaves as a unit. During division the chromosomes become sharply defined, and since in division each chromosome is accurately halved, all the nuclei of the body in general contain the same number of chromosomes. In the "resting-phase" of the nucleus, between two divisions, the chromosomes become indistinct and dispersed through the nucleus. (p. 51.)

Conjugation. The union of two cells accompanied by fusion of their nuclei. (pp. 8, 9.)

Dominant. When two germ-cells unite in fertilization, one of which bears one hereditary character and the other another, if the character borne by one appears in the offspring to the exclusion of that borne by the other, it is said to be dominant.

Factor. The hypothetical unit in a germ-cell which determines the production in the offspring of a particular character or sex.

Fertilization. The conjugation of two germ-cells, usually of a spermatozoon with an ovum; or more generally, the process which stimulates an ovum to begin to develop into a new individual. (pp. 8–10.)

Germ-cell. A reproductive cell, which, usually after union with a germ-cell derived from another individual (conjugation), develops into a new individual. *See* ovum, spermatozoon.

Gynandromorph. An individual of which one part shows the characters of the male, another part those of the female. Most frequently gynandromorphs are *bilateral*, *i.e.* the right side of the body has the characters of one sex, the left those of the other sex. In many cases the arrangement is irregular. (p. 126.)

Heterochromosome or Heterotropic chromosome. A chromosome which is unpaired, so that at one of the cell-divisions by which the germ-cells are produced, it goes undivided into one of the daughter-cells, with the result that half the germ-cells contain it and half do not. Also called "accessory" chromosome. An animal which has such an unpaired heterochromosome in one sex, has a pair of similar chromosomes in the other sex. (pp. 61, 62.)

Heterozygous. An individual which receives an inherited character from one parent but not from the other is said to be heterozygous in respect of that character.

Homozygous. An individual which inherits the same character from both parents is homozygous in respect of that character.

Hormone. A substance produced by some organ (usually ovary or testis) from which it is supposed to diffuse into the blood, and by its circulation with the blood to all parts of the body, to stimulate the production of certain characters. (p. 95.)

Idiochromosome. A chromosome which differs from a heterochromosome in being unequally paired instead of unpaired, in one sex, usually the male. In the production of the spermatozoa, the larger idiochromosome goes into one daughter-cell, the smaller into the other, so that half the spermatozoa have the larger, half the smaller. (p. 63.)

Interstitial Tissue of the testis or ovary, is that part of the organ from which germ-cells are not produced, but which is supposed to produce secretions (hormones) which diffuse into the blood. (p. 95.)

Maturation Divisions (or simply Maturation) are the nuclear divisions of the germ-cells at which the number of chromosomes is reduced from the double complement characteristic of most cells of the body to the single complement characteristic of the mature germ-cell. (pp. 52, 54.)

Metabolism. The sum of the physiological processes connected with nutrition; *i.e.* both the building up of relatively simple food-stuffs into the more complex living substance, and the breaking down of living matter in the performance of its functions into simpler waste-products which are excreted.

Mutation. The sudden origin of a new variety, usually by the disappearance of some character which is present in the typical form.

Glossary of Technical Terms 161

Nucleus. The body present in every cell which appears to control the life-processes of the cell. (pp. 9, 50.)

Ovary. The organ in which ova or eggs are produced.

Ovum. The germ-cell produced by the female, an egg-cell; an egg apart from such extrinsic portions as the shell or other external covering. (pp. 7, 8.)

Parthenogenesis. The production of young from an ovum which is not fertilized by a spermatozoon.

Polar Bodies, Polar Nuclei. The nuclei which are thrown off from the ovum in the process of maturation. (p. 54.)

Protoplasm. The substance of which all living matter ultimately consists. (p. 8, note.)

Protozoa. The simplest animals, which consist of one cell, or of a small amount of protoplasm containing nuclei, but not divided into cells. (p. 11.)

Recessive. When two germ-cells unite in fertilization, one of which bears one hereditary character and the other another, if the character borne by one appears in the offspring to the exclusion of that borne by the other, the one which does not appear is said to be recessive, cf. *Dominant*.

Secondary Sexual Characters. Characters not immediately concerned with reproduction which are found in one sex but not in the other. (p. 91.)

Sex-limited Inheritance. That kind of inheritance in which a character is transmitted by individuals of one sex almost exclusively to offspring of the other sex. (pp. 31, 36.)

Spermatozoon. The germ-cell produced by the male. Spermatozoa are usually very minute and have the power of active locomotion. (p. 8.)

Testis. The organ in which spermatozoa are produced.

Tissue. A mass of cells all of which are modified to perform a common function, *e.g.* muscle, skin, gland.

"X-chromosome." A chromosome which is supposed to have a sex-determining function, in the sense that when two X-chromosomes are present in a fertilized egg, the egg develops into a female; when only one, into a male. (p. 61.)

Zygote. The product of conjugation of two germ-cells; a fertilized ovum. (p. 9.)

REFERENCES

Arkell, T. R. and Davenport, C. B. "Horns in Sheep as a typical sex-limited character." *Science*, XXXV. (1912), p. 375.

Bateson, W., and Punnett, R. C. "The Inheritance of the peculiar pigmentation of the Silky Fowl." *Journ. of Genetics*, I. (1911), p. 185.

—— and Saunders, E. R. *Reports Evol. Comm. Roy. Soc.* I. (1902), p. 138.

Bond, C. J. "On a case of unilateral development of secondary male characters in a Pheasant." *Journ. of Genetics*, III. (1914), p. 205.

Boring, A. M. "Interstitial Cells and the supposed Internal Secretion of the Chicken Testis." *Biol. Bull.* XXIII. (1912), p. 141.

Boveri, T. "Über das Verhalten der Geschlechtschromosomen, etc." *Verh. d. Phys.-Med. Ges. Würzburg*, XLI. (1911), p. 83. (Summarized by Castle, *Amer. Nat.* XLV. (1911), p. 425.)

—— on Gynandromorphs, see Morgan, *Heredity and Sex* (1913), p. 164.

Brauer, A. "Zur Kenntniss der Spermatogenese von *Ascaris megalocephala*." *Arch. Mikr. Anat.* XLII. (1892).

Bridges, C. B. "Non-disjunction of the Sex-chromosomes in *Drosophila*." *Journ. Exp. Zool.* XV. (1913), p. 587.

Bulloch, W., and Fildes, P. "Haemophilia. Treasury of Human Inheritance V. and VI." (*Eugenics Lab. Memoirs*, XII. London, 1911.)

Castle, W. E. "Heredity of Sex." *Bull. Mus. Comp. Zool. Harvard*, XL. (1903), p. 189.

Coe, W. R. "The Maturation and Fertilization of the Egg of *Cerebratulus*." *Zool. Jahrb.* (*Anat.*) XII. (1899).

Dawson, E. R. *The Causation of Sex.* Lewis, London, 1909.

Denso. "Palaearkt. Schmetterlingsformen." *Deutsch. Entom. Zeitschr.* (*Iris*), XXVI. (1912), p. 127.

References

Doncaster, L. "Gametogenesis of the Gall-fly *Neuroterus lenticularis.*" *Proc. Roy. Soc.* B 82, (1910), p. 88 and B 83, (1911), p. 476.

―― "On the relation between chromosomes, sex-limited transmission, and sex-determination in *Abraxas grossulariata.*" *Journ. of Genetics*, IV. (1914), p. 1.

―― "On an inherited tendency to produce purely female families, etc." *Journ. of Genetics*, III. (1913), p. 1.

―― "On the chromosomes...of *Biston hirtaria, Nyssia zonaria,* and their hybrids." *Journ. of Genetics*, III. (1914), p. 234.

―― *Heredity in the Light of Recent Research.* 2nd edition. Camb. Univ. Press, 1912.

―― "On Sex-limited Inheritance in the Cat." *Journ. of Genetics*, III. (1913), p. 11.

―― "Chromosomes, Heredity, and Sex." *Quart. Journ. Micr. Sci* LIX. (1914), p. 487.

Durham, F. M., and Marryat, D. C. E. "Note on the Inheritance of Sex in Canaries." *Evol. Comm. of Royal Soc. Report,* IV. (1908), p. 57.

Dziurzynski, K. "*Bupalus piniarius.*" *Berlin Entom. Zeitschr.* LVII. (1912), p. 1. (Pl. I., Fig. 7.)

Ewart, R. J. "Sex-relationship." *Nature,* LXXXV. (1911), p. 322.

Federley, H. "Sur un cas d'Hérédité gynephore." *IV. Conf. de Génétique,* Paris (1911), p. 469.

Foot, K., and Strobell, E. C. "Preliminary note on the crossing of two Hemipterous species." *Biol. Bull.* XXIV. (1913), p. 187.

Fryer, J. C. F. "An Investigation by pedigree breeding into the polymorphism of *Papilio polytes.*" *Phil. Trans. Roy. Soc.* B 204 (1913), p. 227.

Gerould, J. H. "The Inheritance of polymorphism and sex in *Colias philodice.*" *Amer. Nat.* XLV. (1911), p. 257.

Goldschmidt, R. "Untersuchungen über die Vererbung der sekundären Geschlechtscharaktere." *Zeitschr. f. indukt. Abstamm.* VII. (1912), p. 1.

―― and Poppelbaum, H. "Erblichkeitsstudien an Schmetterlingen. II." *Zeitschr. f. indukt. Abstamm.* XI. (1914), p. 280.

Goodale, H. D. "Some Results of Castration in Ducks." *Biol. Bull.* XX. (1910), p. 35.

Gutherz, S. "Über ein bemerkenswertes Strukturelement, etc." *Arch. Mikr. Anat.* 79 (II.), (1912), p. 79.

Guyer, M. F. "Accessory Chromosomes in Man." *Biol. Bull* XIX. (1910), p. 219.

Hagedoorn, A. L. "Mendelian Inheritance of Sex." *Arch. Entwick-Mech.* XXVIII. (1909), p. 1.

Harrison, J. W. H. "The Hybrid Bistoninae." Oberthür's *Etudes de Lepidopterologie comparée.* Fasc. VII. 1913.

Heape, W. "Note on the influence of extraneous forces upon the proportion of the Sexes produced by Canaries." *Proc. Camb. Philosoph. Soc.* XIV. (1907), p. 201.

—— "Notes on the proportion of the Sexes in Dogs." *Proc. Camb. Philosoph. Soc.* XIV. (1907), p. 121.

—— "The proportion of the sexes produced by Whites and Coloured peoples in Cuba." *Phil. Trans. Roy. Soc.* B 200, (1907), p. 27, (and *Proc. Roy. Soc.* B 81 (1909), p. 32).

Hertwig, R. "Über den derzeitigen Stand des Sexualitäts-problems." *Biol. Zentralbl.* XXXII. (1912), p. 1.

Issakowitsch, A. "Geschlechtsbestimmende Ursachen bei den Daphniden." *Biol. Centralbl.* XXV. (1905), p. 529; *Arch. Mikr. Anat.* LXIX. (1906–7), p. 223.

Jordan, H. E. "The Spermatogenesis of the Opossum." *Arch. Zellforsch.* VII. (1911), p. 41.

King, H. D. "Studies in Sex-determination of Amphibians." *Biol. Bull.* XX. (1911), and *Journ. Exp. Zool.* XII. (1912), p. 319.

—— "The Sex-ratio in Hybrid Rats." *Biol. Bull.* XXI. (1911). p. 104.

Kopeč, S. "Untersuchungen über Kastration, etc." *Arch. Entwick-Mech.* XXXIII. (1911), p. 1.

Krüger, E. "Die Phylogenetische Entwicklung der Keimzellen einer freilebenden Rhabditis," *Zool. Anzeiger*, XL. (1912), p. 233; and *Zeitschr. wiss. Zool.* CV. (1913), p. 87.

Lang, A. "Vererbungs-wissenschaftliche Miszellen." *Zeitschr. f. indukt. Abstamm.* VIII. (1912), p. 233.

Loeb, J. *Artificial Parthenogenesis and Fertilization.* Chicago, 1913.

Marchal, P. "La polyembryonie spécifique, ou germinogonie." *Arch. Zool. Exp. et Gén.* (4) II. p. 257.

Marshall, F. H. A. *Physiology of Reproduction.* London, 1910.

Marshall, F. H. A. "On the effects of Castration and Ovariotomy upon Sheep." *Proc. Roy. Soc.* B 85 (1912), p. 27.

de Meijere, J. C. H. "Über Jacobsons Zuchtungsversüche, etc." *Zeitschr. f. indukt. Abstamm.* III. (1910), p. 161.

Meisenheimer, J. *Experimentelle studien zur Soma und Geschlechts differenzierung.* Jena, 1909.

Meves, J. F. "Die Spermatocytenteilungen bei der Honigbiene." *Arch. Mikr. Anat.* LXX. (1907), p. 414, and (Hornet) LXXI. (1908), p. 571.

Morgan, T. H. *Heredity and Sex.* Columbia Univ. Press, New York 1913.

[This book includes a general account of Sex-limited inheritance in *Drosophila*, as well as other forms. It contains an excellent bibliography of the whole subject of Sex-determination.]

For *Drosophila*, see also Morgan, *Science*, XXXII. (1910), p. 120; *Amer. Nat.* XLV. (1911), p. 65; *Zeitschr. f. indukt. Abstamm.* VII. (1912), p. 323; *Journ. Exp. Zool.* XI. (1911), p. 365, and XIII. (1912), p. 27, etc.; and Morgan and Cattell, *Journ. Exp. Zool.* XIII. (1912), p. 79.

—— "A biological Study of Sex-determination in Phylloxerans and Aphids." *Journ. Exp. Zool.* VII. (1909), p. 239; see also *Heredity and Sex*, New York, 1913.

—— "The Elimination of the Sex-chromosomes from the male-producing eggs of Phylloxerans." *Journ. Exp. Zool.* XII. (1912), p. 479.

—— "Explanation of a new sex-ratio in Drosophila." *Science*, XXXVI. (1912), p. 718.

Newman, H. H. "Natural History of the Nine-banded Armadillo of Texas." *Amer. Nat.* XLVII. (1913), p. 513; also Newman and Patterson, *Journ. Morphol.* XXI. (1910), p. 359.

Pearl, R., and M. D. "On the relation of race-crossing to the Sex-ratio." *Biol. Bull.* XV. (1908), p. 194.

Pearl, R., and Surface, F. M. "On the Inheritance of the barred pattern in Poultry." *Arch. Entwick-Mech.* XXX. (1910), p. 45.

Pearl, R., and Parshley, H. M. "Data on Sex-determination in Cattle." *Biol. Bull.* XXIV. (1913), p. 205.

Phillips, J. C. "A further study of size-inheritance in Ducks" *Journ. Exp. Zool.* XVI. (1914), p. 143.

Poll, H. " Zur Lehre von den sekundären Geschlechscharakteren.' *Sitzb. d. Ges. Naturf. Freunde Berlin* (1909), p. 338.

Potts, F. A. "Notes on the free-living Nematodes." *Quart. Journ. Micr. Sci.* LV. (1910), p. 433.

Poulton, E. B. "All-female batches of *Acraea encedon*, L., bred in the Lagos district by Mr W. A. Lamborn." *Proc. Entom. Soc.* (1911), p. liv.

Punnett, R. C. *Mendelism*, London, Macmillan, 1911.

—— "On Nutrition and Sex-determination in Man." *Proc. Camb. Philosoph. Soc.* XII. (1903), p. 262.

Reuter, E. "Eibildung bei der Milbe *Pediculopsis graminum*." *Festschrift für Palmén*, vol. I., Helsingfors, 1905–07.

Schenk, L. "Schenk's Theory; The Determination of Sex." London, 1898.

Schleip, W. "Das Verhalten des Chromatins bei *Angiostomum (Rhabdonema) nigrovenosum*." *Arch. Zellforsch.* VII. (1911), p. 87.

Seiler, J. "Das Verhalten der Geschlechtschromosomen bei Lepidopteren. *Zool. Anz.* XLI. (1913), p. 246.

Shattock, S. G., and Seligmann, C. G. "Observations upon the Acquirement of Secondary Sexual Characters, indicating... an internal Secretion by the Testicle." *Proc. Roy. Soc.* B 73 (1904), p. 49.

Shearer, C. "The Problem of Sex-determination in *Dinophilus gyrociliatus*." *Quart. Journ. Micr. Sci.* LVII. (1912), p. 329.

Shull, A. F. "Studies in the Life-cycle of *Hydatina senta*." *Journ. Exp. Zool.* XII. (1912), p. 283.

Silvestri, P. " Contribuzioni alla conoscenza biologica degli Imenotteri parasiti, I. Litomastix. *Atti. R. Acad. Lincei*, XIV. (1905), p. 534, and *Ann. R. Scuola Sup. d'Agricoltura Portici*, VI. 1906.

Smith, Geoffrey. "Studies in the Experimental Analysis of Sex." *Quart. Journ. Micr. Sci.* LV. (1910), p. 225; 57 (1911), p. 251; 59 (1913), p. 267.

—— *Cambridge Natural History*. Vol. "Crustacea and Arachnids." Macmillan, 1909.

—— and Haig Thomas, R. "On Sterile and Hybrid Pheasants." *Journ. of Genetics*, III. (1913), p. 39.

Standfuss, M. [on *Aglia tau* of which one side has *lugens* character]. *Berlin. Ent. Zeitschr.* XXXII. (1888), p. 239.

References

Standfuss, M. "Einige Andeutungen bezüglich der Bedeutung... des sexuellen Färbungs-Dimorphismus." *Mittheil. Schweiz. Entom. Ges.* XII. (1913), p. 99.

Staples-Browne, R. "Second Report on the Inheritance of Colour in Pigeons." *Journ. of Genetics*, II. (1912), p. 131.

Steche, O. "Die sekundären Geschlechtscharaktere der Insekten." *Zeitschr. f. indukt. Abstamm.* VIII. (1912), p. 284.

Steinach, E. "Feminierung von Männchen und Maskulierung von Weibchen. *Zentralbl. f. Physiol.* XXVII. (1913), p. 3.

Stevens, N. M. "Preliminary Note on Heterochromosomes in the Guinea-pig." *Biol. Bull.* XX. (1911), p. 121.

Stichel, H. "Über Melanismus und Nigrismus bei Lepidopteren." *Zeitschr. f. Wiss. Insektenbiol.* VIII. (1912), p. 41.

Thomas, R. Haig. "Experimental Pheasant-breeding." *Proc. Zool. Soc.* III. (1912), p. 539.

Toyama, K. "Studies on Hybridology of Insects, I." *Bull. Coll. Agric. Tokyo Imp. Univ.* VII. 1907.

Whitney, D. D. "The Influence of External Conditions upon the Life-cycle of *Hydatina senta*." *Science*, XXXII. (1910), p. 345.

Wilson, E. B. "Studies on Chromosomes." *Journ. Exp. Zool.* III. (1906), p. 1; VI. (1909), p. 69 and p. 147; *Journ. of Morphol.* XXII. (1911), p. 71.

von Winiwarter, H. "Études sur la spermatogenèse humaine." *Arch. de Biol.* XXVII. (1912), p. 91.

Wood, T. B. "The Inheritance of Horns and Face-colour in Sheep." *Journ. of Agric. Sci.* III. (1909), p. 145.

Wodsedalék, J. E. "Accessory Chromosomes in the Pig." *Science*, XXXVIII. (1913), p. 30; and *Biol. Bull.* XXV. (1913), p. 8.

Zur Strassen. *See* Potts, F. A.

INDEX

Abraxas grossulariata (Plates V, XIV) 22, 31–37, 41, 42, 47, 65, 67, 70, 75 (note), 107, 128, 132, 143
Accessory chromosome 62 (note), 64 (see also '*X*-chromosome')
Acraea encedon 132
Age of parents 79, 81
Ageniaspis 25
Aglia tau 131
Alternation of male- and female-determining ova 153
Amphibia 82–86, 145
Amphidasys betularia (Plate XXI) 128
Ant, gynandromorph (Plate XX) 127
Aphids 19, 20, 21, 59, 71, 76
ARKELL, T. R. and DAVENPORT, C. B. 111
Armadillo (Plate IV) 28

BAEHR, W. B. VON 591
Balance, analogy of in sex-determination 144–146
BALTZER, F. 125 (note)
Barnacles 102, 124
Barred Plymouth Rock (Plate VI) 38
BATESON, W. 138
BATESON, W. and PUNNETT, R. C. 38

Bee, see Honey Bee
Binucleate eggs 128
Birds 15, 22, 29, 32, 37–39, 49, 64, 77, 91, 96, 108, 138, 145
Biston hirtaria (Plate XVI) 110, 132, 133, 142
Blood, differences in two sexes 101, 103
BOND, C. J. 98
Bonellia (Plate I) 5, 92, 125 (note)
Bopyrus 123
BORING, A. M. 97 (note)
BOVERI, T. 127
Bradynema rigidum 122
BRAKE, B. 115
Breeding Season 78, 93
BRIDGES, C. B. 68
Bullfinch, gynandromorph (Frontispiece) 98, 127
Bupalus piniarius (Plate XX) 127
Butterflies, see Lepidoptera

Canary 37, 38, 42, 47, 77
Capon 96
CASTLE, W. E. 138
Castration 94–97, 100
 in Mammals 94
 in Birds 96, 108
 in Insects 100
 parasitic 102–106
Cat 29, 39, 45, 46, 65, 69, 79, 157

Index

Cattle 29, 83, 84, 152
'Cause,' use of word 4, 71
Chaffinch, gynandromorph 98
Chalcididae 25
Chromatin 9, 50, 137
Chromosomes 51–72 (see also 'Sex-chromosomes')
  as bearers of sex-limited characters 67–72, 141
Colias philodice 112
Colour-blindness 43–45, 150
Complemental Males 124
Crab, effects of *Sacculina* on 101–106, 117, 142
Crustacea, hermaphrodite 123–125
Currant Moth, see *Abraxas*
Cynipidae 19, 24, 57

Daphniidae 21, 76
DARWIN, C. 124
DAVENPORT, C. B. 39
DAWSON, E. RUMLEY 153
Deer 94
Determination of Sex, meaning of term 3, 4
  period of 16, 17, 88, 136
  Mendelian theories of 118, 138, 140
Dimorphism, see Sexual Dimorphism
  of eggs 18–22
  of spermatozoa 29, 30, 61–65, 75 (note)
Dinophilus 21
Dog 74, 78, 83, 152
Doves 38, 47
Drone 23, 57
Drosophila 39–45, 47, 65, 68, 81 (note), 131, 141
Duck 15, 86, 87, 97
DURHAM, F. M. 38
DZIERZON 23

Egg, male-producing or female-producing 18–22, 35, 36
Egg-production, inheritance of 39
Encyrtus (Plate III) 25
Euschistus 110
EWART, R. J. 79

FEDERLEY, H. 131
Fertilization (Plate XI) 8, 10–12
FOOT, K. and STROBELL, E. C. 110
Fowls 22, 38, 39, 42, 96, 108
"Free-martin" 29
Frog 82–86, 88, 139, 143, 149, 151
FRYER, J. C. F. 113

Gall-flies 19, 24, 57, 60, 71
Germ-cells, maturation of 52–66
GEROULD, J. H. 112
GOLDSCHMIDT, R. 115, 133
  and POPPELBAUM, H. 115 (note)
GOODALE, H. D. 39, 97
Grafting of ovaries and testes 95, 100, 142
Guinea-pig 95, 150
GUTHERZ, S. 150
GUYER, M. F. 149, 150 (note)
Gynandromorph Birds 98, 127
  Insects 115, 126
Gynandromorphism (Plates XX, XXI, XXII, XVIII, Frontispiece) 126–131, 133

Haemophilia 43–45, 47, 48, 150, 156
HAGEDOORN, A. L. 39
Hag-fish 125
HAIG-THOMAS, R. 86, 109
HARRISON, J. W. H. 133
HEAPE, W. 77–79, 149

Index

HEGNER, R. W. 26 (note)
Hemiptera 62
Hermaphroditism 3, 119–126
HERTWIG, R. 82–86, 88, 116 (note), 125, 139, 144, 149, 151
Heterochromosome 62 (note)
Heterotropic chromosome (Plate XIII) 62 (note), 64
Honey Bee 23, 24, 56, 60, 127, 128
Hormones 95, 97, 98, 105, 110
Horns of Sheep 94, 111
Horse 74, 80, 83
Hybrids, Sex-ratio in 86
Hydatina (Plate I) 18, 77 (see also Rotifers)
Hymenoptera 19, 23–27, 58, 60

Idiochromosome (Plate XIII) 63
Inachus 101, 123
'Indifferent' frog-larvae 85, 125, 144, 152
Insects 18–20, 23–27, 32–37, 39–44, 49, 57–69, 91, 98, 100, 116, 138, 143, 144, 145
Interstitial Tissue 95–97
ISSAKOWITSCH, A. 76

JACOBSON, E. 114 (note)
JORDAN, H. E. 150

KING, H. D. 83, 151
KOPEČ, S. 100
KRÜGER, E. 122
KUSCHAKEWITSCH, A. 82

Lepidoptera 5, 15, 22, 32–37, 64–68, 86, 138
(see also *Abraxas*)
Litomastix (Plate III) 25–27
LOEB, J. 151

Lymantria dispar 100
hybrids with *L. japonica* 115, 132
Lymantria monacha 131

Mammals 28–30, 43–49, 64, 83, 91, 94, 116, 138, 142, 145, 148, 150, 152, 158
Man 27, 29, 39, 43–45, 47, 48, 64, 65, 74, 78–81, 87, 139, 147–158
MARCHAL, P. 25
MARSHALL, F. H. A. 94
Material basis of sex-determination 50–72
Maturation of germ-cells (Plates IX, X) 52–66
MEIJERE, J. C. H. DE 114 (note)
MEISENHEIMER, J. 100
Mendelian theories of sex-determination 118, 138, 140
Molluscs, hermaphrodite and bisexual 120
Monthly alternation of male and female ova 153
MORGAN, T. H. 39–42, 47, 59, 81 (note), 127, 128 (note), 131
Moths, see Lepidoptera
Mouse 87, 112 (note)
Myriapoda 64, 138
Myxine 125

Nematoda 64, 120–123, 138, 142
NEWMAN, H. H. 28
Night-blindness 43–45, 150, 156
Nucleus 8–12, 50
division of (Plate XI) 10, 51
Nutrition 76, 154
Nyssia zonaria (Plate XVI) 110, 132, 134, 142
Nystagmus 43–45

Index

Opossum 150
Orchestia 124
Orgyia 5
Ornithoptera 15
Orthoptera 61
Ovary, removal of 94–97, 153
Ovum 7
   male-producing or female-producing 18–22, 35, 36, 57–59, 65–67, 79, 136, 153–155
   maturation of (Plates X, XIV) 54

Papilio 37, 112
   dardanus 114 (note)
   memnon 114 (note)
   polytes (Plate XVII) 113, 117
Paramoecium 11, 12
Parasitic castration 102–106
Parnassius delius (Plate XX) 127
Parthenogenesis 6, 18–21, 24, 50, 58, 59, 61, 77, 122
PEARL, R. 39, 83, 85, 87, 152
PEARL, R. and SURFACE, F. M. 38
Pediculopsis 22
Pheasant 15, 86, 109
   gynandromorph 98, 127
   Formosan 86, 109
   Reeve's 86
   Silver 86, 109
   Swinhoe 86, 109
   Versicolor 86, 109
PHILLIPS, J. C. 86, 87
Phragmatobia fuliginosa (Plate XIV) 65
Phyllopoda (see also *Daphniidae*) 59
Phylloxera 18, 59
Physiological differences between the sexes 13–15, 90, 101, 103, 116, 132, 137, 144

Pig 150
Pigeons 38, 42
Poecilopsis pomonaria 133
Polar bodies 54
POLL, H. 97 (note), 98
Polyembryony 25–29
POULTON, E. B. 132
Proctotrypidae 25
Protozoa 11, 12
Psychidae 5
PUNNETT, R. C. 80–81
Pygaera pigra 131

Rat 74, 87, 95, 153
RAYNOR, G. H. 32 (note)
REUTER, E. 22
Rhabditis aberrans 122
Rhabdonema nigrovenosum (Plate XIX) 120, 142
Right and left ovaries, supposed to produce different sexes 153
Rotifers 18–21, 59, 76

Sacculina (Plate XV) 101–106, 117, 142
Saw-flies 25
Scalpellum 124
SCHENK, L. 154
Secondary sexual characters 5, 15, 31, 90–118
   inheritance of 107–118
   transmission of, by one sex to the other 108–110
SEILER, J. 65
Selective fertilization, hypothesis of 140, 155
Sex
   nature of 1–3, 7–14
   function of 1–3, 7–13
Sex-chromosomes 62–72, 126, 133, 135, 138, 142–145, 152
   (see 'X-chromosome')

Sex-limited inheritance 22, 29, 31–49, 65–70, 107, 113, 136, 139, 142, 150, 156–158
  exceptions 35–37, 38, 46, 47, 68
Sex-ratio 5, 73–89, 151
  abnormal 47–48, 82–88, 131–133, 156
  variations of 74, 82–88, 139, 152
Sexual Dimorphism 5, 92
Sexual Distinctions 8, 13–15, 90–92, 101, 116
SHEARER, C. 21
Sheep 94, 111
SHULL, A. F. 76
Silk-worm (Plate XXII) 129
Silky Fowl 38
SILVESTRI, F. 25
Simocephalus 76
SMITH, GEOFFREY 101–106, 123, 124
Spermatozoon 8
  male- or female-producing 29, 30, 42, 49, 62–69, 137, 139, 150
  development of (Plates IX, XII) 53–57
Spider 22, 64, 138
Staleness of eggs, effects on sex-ratio 82–84, 139
STANDFUSS, O. 101 (note), 131
STAPLES-BROWNE, R. 38
STECHE, O. 100, 101, 104
STEINACH, E. 95, 99, 108, 142

STEVENS, N. M, 150
STICHEL, H. 131

Tatusia novemcincta (Plate IV) 28
Temperature 76, 77, 86, 101 (note)
Tenthredinidae 25
Toad 83
TOYAMA, K. 129
Transplantation of ovaries and testes 95, 100, 142
Turtle dove 38, 47
Twins 27–29, 153
Unisexual families 131–133

Vertebrates, hermaphrodite 125

"Water-fleas" 21, 59, 76
WEBER 98
WHITNEY, D. D. 76
WILSON, E. B. 63
WINIWARTER, H. VON 69, 139, 149, 150 (note)
WODSEDALÉK, J. E. 150
WOOD, T. B. 111

X-chromosome' 59–70, 83–85, 112, 120, 149

'Y-chromosome' 63
Yolk-forming substances 103–105

ZUR STRASSEN 123
Zygote 9

www.ingramcontent.com/pod-product-compliance
Ingram Content Group UK Ltd.
Pitfield, Milton Keynes, MK11 3LW, UK
UKHW050346180125
453697UK00016B/453